Badredine Arfi

Linguistic Fuzzy Logic Methods in Social Sciences

Studies in Fuzziness and Soft Computing, Volume 253

Editor-in-Chief

Prof. Janusz Kacprzyk
Systems Research Institute
Polish Academy of Sciences
ul. Newelska 6
01-447 Warsaw
Poland
E-mail: kacprzyk@ibspan.waw.pl

Further volumes of this series can be found on our homepage: springer.com

Vol. 237. Kofi Kissi Dompere
Fuzziness and Approximate Reasoning, 2009
ISBN 978-3-540-88086-8

Vol. 238. Atanu Sengupta, Tapan Kumar Pal
Fuzzy Preference Ordering of Interval Numbers in Decision Problems, 2009
ISBN 978-3-540-89914-3

Vol. 239. Baoding Liu
Theory and Practice of Uncertain Programming, 2009
ISBN 978-3-540-89483-4

Vol. 240. Asli Celikyilmaz, I. Burhan Türksen
Modeling Uncertainty with Fuzzy Logic, 2009
ISBN 978-3-540-89923-5

Vol. 241. Jacek Kluska
Analytical Methods in Fuzzy Modeling and Control, 2009
ISBN 978-3-540-89926-6

Vol. 242. Yaochu Jin, Lipo Wang
Fuzzy Systems in Bioinformatics and Computational Biology, 2009
ISBN 978-3-540-89967-9

Vol. 243. Rudolf Seising (Ed.)
Views on Fuzzy Sets and Systems from Different Perspectives, 2009
ISBN 978-3-540-93801-9

Vol. 244. Xiaodong Liu and Witold Pedrycz
Axiomatic Fuzzy Set Theory and Its Applications, 2009
ISBN 978-3-642-00401-8

Vol. 245. Xuzhu Wang, Da Ruan, Etienne E. Kerre
Mathematics of Fuzziness – Basic Issues, 2009
ISBN 978-3-540-78310-7

Vol. 246. Piedad Brox, Iluminada Castillo, Santiago Sánchez Solano
Fuzzy Logic-Based Algorithms for Video De-Interlacing, 2010
ISBN 978-3-642-10694-1

Vol. 247. Michael Glykas
Fuzzy Cognitive Maps, 2010
ISBN 978-3-642-03219-6

Vol. 248. Bing-Yuan Cao
Optimal Models and Methods with Fuzzy Quantities, 2010
ISBN 978-3-642-10710-8

Vol. 249. Bernadette Bouchon-Meunier, Luis Magdalena, Manuel Ojeda-Aciego, José-Luis Verdegay, Ronald R. Yager (Eds.)
Foundations of Reasoning under Uncertainty, 2010
ISBN 978-3-642-10726-9

Vol. 250. Xiaoxia Huang
Portfolio Analysis, 2010
ISBN 978-3-642-11213-3

Vol. 251. George A. Anastassiou
Fuzzy Mathematics: Approximation Theory, 2010
ISBN 978-3-642-11219-5

Vol. 252. Cengiz Kahraman, Mesut Yavuz (Eds.)
Production Engineering and Management under Fuzziness, 2010
ISBN 978-3-642-12051-0

Vol. 253. Badredine Arfi
Linguistic Fuzzy Logic Methods in Social Sciences, 2010
ISBN 978-3-642-13342-8

Badredine Arfi

Linguistic Fuzzy Logic Methods in Social Sciences

Springer

Author

Prof. Badredine Arfi
Department of Political Science
University of Florida
234 Anderson Hall, POB 117325
Gainesville, FL 32611
USA
E-mail: barfi@ufl.edu

ISBN 978-3-642-42237-9 ISBN 978-3-642-13343-5 (eBook)

DOI 10.1007/978-3-642-13343-5

Studies in Fuzziness and Soft Computing ISSN 1434-9922

Softcover re-print of the Hardcover 1st edition 2010

Typeset & Cover Design: Scientific Publishing Services Pvt. Ltd., Chennai, India.

Printed on acid-free paper

9 8 7 6 5 4 3 2 1

springer.com

First Foreword

The modern origin of fuzzy sets, fuzzy algebra, fuzzy decision making, and "computing with words" is conventionally traced to Lotfi Zadeh's publication in 1965 of his path-breaking refutation of binary set theory. In a sixteen-page article, modestly titled "Fuzzy Sets" and published in the journal Information and Control, Zadeh launched a multi-disciplinary revolution. The start was relatively slow, but momentum gathered quickly. From 1970 to 1979 there were about 500 journal publications with the word fuzzy in the title; from 2000 to 2009 there were more than 35,000. At present, citations to Zadeh's publications are running at a rate of about 1,500-2,000 per year, and this rate continues to rise.

Almost all applications of Zadeh's ideas have been in highly technical scientific fields, not in the social sciences. Zadeh was surprised by this development. In a personal note he states: "When I wrote my 1965 paper, I expected that fuzzy set theory would be applied primarily in the realm of human sciences. Contrary to my expectation, fuzzy set theory and fuzzy logic are applied in the main in physical and engineering sciences." In fact, the first comprehensive examination of fuzzy sets by a social scientist did not appear until 1987, a full twenty-two years after the publication of Zadeh's seminal article, when Michael Smithson, an Australian psychologist, published *Fuzzy Set Analysis for Behavioral and Social Sciences.* Smithson explored many different aspects of fuzzy set theory, with a special focus on points of convergence between fuzzy set theory and conventional quantitative analysis. However, this work did not attract the following it deserved, and few scholars took up his call to explore the possibilities offered by this new approach.

The 1987 also saw the publication of my book *The Comparative Method: Moving Beyond Qualitative and Quantitative Strategies*. In this work I focused on conventional Boolean sets and explored the use of set-theoretic methods for studying cases as configurations of aspects and for comparing cases as configurations. In contrast to Smithson's book, my work emphasized the distinctiveness of set-theoretic methods and the gap between these methods and conventional quantitative methods. However, the techniques I developed were limited to conventional Boolean sets, and many social scientists were dismayed at the thought of using dichotomies, especially given their deep commitments to interval-scale measures and multivariate techniques such as multiple regression.

For these and other reasons, I took up Smithson's call to explore fuzzy sets, which culminated in 2000 with the publication of my book *Fuzzy-Set Social Science*. I extended and elaborated these ideas in 2008 with the publication of

Redesigning Social Inquiry: Fuzzy Sets and Beyond. These two books join the configurational methods I present in my 1987 book with the fuzzy set principles developed by Zadeh and his many followers. The tools I present provide methods for comparing cases as configurations, and they allow for the possibility that cases may have only partial memberships in the sets that comprise a given configuration. In effect, a configuration may be seen as an ideal type, as Weber would describe it, with cases varying in the degree to which they conform to that ideal type. By looking at cases as configurations, it is possible to explore causal heterogeneity and causal complexity in ways that are beyond the grasp of conventional quantitative methods.

In both the 2000 and 2008 books, I use fuzzy sets that are numerically based. That is, each case is assigned a numerical value between zero (0) and one (1.0) to signal its degree of membership in each relevant set. The conceptualization and labeling of each set used in an investigation has a determining impact on the assignment of numerical values. For example, a survey respondent might receive a score of 1.0 for his or her membership in the set of "at least moderate" income individuals, but a score of only 0.60 in the set of "high" income individuals. The key idea is that different conceptualizations of the target set call for different "calibrations" of set membership, even when using the same source variable (e.g., income, as conventionally measured). In this way, researchers can establish a close correspondence between the concepts they use in their verbal arguments and their practical analytic procedures. This close coupling of concepts and calibrations contrasts sharply with conventional approaches in quantitative work, where there is heavy reliance on "indicators" that simply "correlate" with underlying concepts.

Most applications of fuzzy sets in the social sciences today use numerical values to represent degree of membership in sets. While convenient, this practice does not exhaust the many possibilities offered by fuzzy set theory. An alternate approach, which Badredine Arfi uses to great advantage in the present work, utilizes the actual verbal labels that apply to the evidence, instead of numerical values, when representing degree of membership in sets. This "linguistic" approach to fuzzy sets is the foundation for "computing with words" instead of numbers and also a major focus of what is known today as "soft computing." As Arfi shows, it is possible to establish a very tight connection between verbal theory and argumentation, on the one hand, and empirical analysis, on the other, using linguistic fuzzy logic methods. Arfi's approach, in effect, offers the possibility of a one-to-one correspondence between verbal argumentation and empirical analysis. His approach also eliminates the intermediate task of assigning numerical values to verbal labels.

Arfi's linguistic approach is not the only distinguishing feature of this innovative work. Most applications of fuzzy set methods, especially in the social sciences and elsewhere, use what are known as "type-1" fuzzy sets. That is, specific linguistic labels or numerical values are assigned to cases, indicating the degree of membership of each case in each set. Arfi, by contrast, utilizes "type-2" fuzzy sets, enriching fuzzy set analysis with an additional dimension. A type-2 fuzzy set includes not only some indication of degree of membership (i.e., the type-1 designation), but also an evaluation (which is linguistic in Arfi's work) of the degree of confidence (or truthfulness) that exists in the assigned membership. For example, a researcher might report that a case has "very high" membership in a given set, but also indicate that there is only a moderate degree of confidence in that assignment. Using type-2 fuzzy sets, it is possible, in effect, to skew the analysis so that it favors the evidence in which there is more confidence, and in Arfi's work this can be accomplished using non-numerical (i.e., linguistic) assessments of both set membership and degree of confidence in the assignment of membership.

Arfi demonstrates the analytic power of this combination of linguistic fuzzy logic methods and type-2 fuzzy sets in this pioneering book. After laying the analytic foundations early on, he applies his approach to a range of problems familiar to social scientists, from fuzzy decision making, fuzzy game theory and strategic interaction to fuzzy propositional logic (applied to causality) and fuzzy data analysis. Social scientists who wish to extend their thinking about the possibilities offered by fuzzy sets will find a wealth of new material in this work.

Charles C. Ragin
Department of Sociology
University of Arizona
Tucson, Arizona 85721
USA

Second Foreword

It is hard to add substantively to the foreword of Professor Ragin and the preface of Professor Arfi. Both are informative, authoritative and erudite. In the following, I will confine myself to a few observations.

Allow me to start with a truism. Science deals not with reality but with models of reality. In large measure, scientific progress is driven by a quest for better models of reality.

Science has always been and continues to be based on bivalent logic—a logic in which no shades of truth are allowed. Correspondingly, in fuzzy logic classes are assumed to have sharply defined boundaries. The brilliant successes of bivalent-logic-based science are visible to all. However, the brilliant successes are far more visible in the world of physical sciences than in the world of human sciences—sciences such as sociology, psychology, political science, linguistics and philosophy. There is a reason. The world of physical sciences is the world of measurements, while the world of human sciences is the world of perceptions. Perceptions are intrinsically imprecise, reflecting the bounded ability of human sensory organs and ultimately the brain, to resolve detail and store information. As a consequence, in the world of human sciences classes have unsharp boundaries. Bivalent logic is intolerant of imprecision and partiality of truth. For this reason, bivalent logic is not the right logic for serving as a foundation for human sciences. What is needed for this purpose is fuzzy logic. Essentially, fuzzy logic is the logic of classes with unsharp boundaries. Equivalently, fuzzy logic may be viewed as a precise logic of imprecision. At this juncture, the view that fuzzy logic has much to offer to human sciences has not gained wide acceptance. Eventually, it will, thanks to the pioneering contributions of Smithson, Ragin, Arfi, Mordeson and his collaborators, and others.

In sum, fuzzy logic is needed for dealing with imprecision—especially with unsharpness of class boundaries. But somewhat paradoxically, most applications of fuzzy logic are in the domain of engineering systems—systems in which models are measurement-based and precise. Fuzzy logic is employed extensively in digital cameras, washing machines, air-conditioning systems, automobile transmissions, industrial control and many other application areas. There is an interesting explanation. Precision carries a cost. In most applications there is some tolerance for imprecision. This tolerance is exploited through deliberate imprecisiation of a precisely-defined system. Then the machinery of fuzzy logic is employed to deal with the imprecision of the resulting system. Typically, the resulting system is a linguistic summary of the original system.

Generally, imprecisiation involves the use of linguistic variables and fuzzy if-then rules. These concepts were introduced in my 1973 paper "Outline of a New Approach to the Analysis of Complex Systems and Decision Processes." The concepts of linguistic variable and fuzzy if-then rules served as a basis for the development of linguistic approaches to systems analysis. A case in point is the Linguistic Fuzzy Logic of Professor Arfi. In Professor Arfi's logic, grades of membership and truth values are allowed to be linguistic variables. Professor Arfi's approach opens the door to construction of better models of reality in the realm of human sciences. His work is an important contribution.

In conclusion, Professor Arfi's book, "Linguistic Fuzzy Logic Methods in Social Science," contains much that is new and original. His work paves the way for important applications of fuzzy logic in the realm of human sciences. Professor Arfi and the publisher, Springer, deserve our thanks and congratulations.

Lotfi A. Zadeh
Director, Berkeley Initiative in Soft Computing
Department of Electrical Engineering and Computer Sciences
University of California, Berkeley, CA 94720-1776, USA
May 27, 2010

Preface

Many years ago Charles S. Pierce stated that "vagueness is no more to be done away with in the world of logic than friction in mechanics." Likewise, Bertrand Russell retorted, "everything is vague to a degree you do not realize till you have tried to make it precise." Not surprisingly, vagueness and equivocation are constitutive features of human systems of communication. Human beings cannot live and communicate without using natural languages, yet the latter are inherently vague and equivocal. Addressing the problem of linguistic vagueness is no easy task because most social sciences variables are often difficult to precisely operationalize, a problem that confronts equally those inclined to use either qualitative or quantitative methods of analysis. Constantly seeking more conceptual and operational precision and crispness is hence commonly believed to be a golden rule to the production of high quality research work in social sciences.

Contra this conventional wisdom, this book raises the question: *What if instead of seeking to get rid of language-caused vagueness we make its preservation an essential requirement for analyzing social science phenomena, both theoretically and empirically*? This book develops a novel approach, termed as linguistic fuzzy-logic methodology, which in essence addresses the problem of language-related vagueness in analyzing social sciences phenomena. Exploring the relatively new field of fuzzy-logic as applied to social and political phenomena takes us into a yet to be explored rich territory of research questions, theoretical arguments, and policy insights. Every theory and every empirical analysis of social phenomena is underpinned in a given logic of reasoning for judging coherence, consistency and the possibility of falsification. Much of what counts as social science today is based on a Boolean logic with two truth values, 0 and 1. The task set out for this book is to go beyond this assumption by showing that linguistic fuzzy logic can be very useful in analyzing empirical data, studying causation, and strategic reasoning in social science theories.

The book, titled "Linguistic Fuzzy-Logic Methods in Social Sciences," is a first in its kind. Linguistic fuzzy logic theory deals with sets or categories whose boundaries are blurry or, in other words, "fuzzy," and which are expressed in a formalism that uses "words" to compute, not numbers, termed in engineering as "soft computing." This book presents an accessible introduction to this linguistic fuzzy logic methodology, focusing on its applicability to social sciences. Specifically, this is the first book to propose an approach based on linguistic fuzzy-logic and the method of computing with words to the analysis of decision making processes, strategic interactions, causality, and data analysis in social sciences. The project consists of systematic, theoretical and practical discussions and developments of these new methods as well as their applications to various substantive issues of interest to international relations scholars, political scientists, and social scientists in general.

The book, first, addresses the basic concepts of the methodology. Linguistic fuzzy logic methodology is an analytic framework for handling concepts that are categorical and coded in words, not in numbers. The book begins with a rationale for linguistic fuzzy logic and then introduces readers with a vey elementary requirement of statistics and formal modeling to the necessary concepts and techniques of linguistic fuzzy logic methodology. Second, the book presents novel ways of analyses in social sciences. Readers are shown new methods for testing theories that are expressed in linguistic terms. Issues of operationalizing and testing models expressed in linguistic fuzzy-logic framework are a few of the topics addressed. Third, the book presents various illustrations of the techniques and applications of the new methodology. Concrete examples and data-sets from various branches of the social sciences are used to compare and contrast the linguistic fuzzy logic framework with other analytic techniques. Empirical applications of the technique as well as critiques of fuzzy set theory are presented.

The book has several chapters, successively dealing the process of fuzzy decision making, fuzzy strategic interaction and fuzzy game theory, fuzzy causality and fuzzy propositional logic, and fuzzy data analysis. This manuscript thus presents novel ideas as no other book in social sciences or elsewhere has ever done before. It makes a strong case through its scope and rigor for adopting linguistic fuzzy logic, thereby introducing the subject-area of "computing with words" into social sciences methods. The book specifically plows new grounds by presenting a new kind of game theory and a new type of data analysis methodology.

Given its scope and breadth in covering decision making theory, strategic interaction models, data analysis method, and the formal study of causality, while requiring minimal mathematical, statistical, or computer skills, the book should appeal to a diverse audience: students at the advanced college and university level, graduate students and scholars who are interested in new methods of empirical analysis and theory falsification across all social science disciplines. The major instructor problem in teaching the content of this book will be more at the conceptual level because this methodology goes against the widely held misperception that we need "numbers" and related logic to achieve more rigor and accuracy in studying social and political phenomena in a scientific way. Yet this is not too difficult a challenge since linguistic fuzzy logic methodology is a more or less reflection of our "natural" way of thinking and communicating via natural human language.

Acknowledgements

Over the long and sometimes lonely course of writing this book I have undoubtedly acquired a number of significant debts. I thank all those who helped me in one way or another, even if I do not mention them by name and even if they do not know that they helped me.

The seminal idea for this book emerged at the University of Illinois at Urbana-Champaign, more or less began to consolidate at Southern Illinois University at Carbondale, and finally came into fruition as a book manuscript at the University of Florida at Gainesville. I am grateful for the time and support provided by all three institutions.

I thank both Prof. Garry Goertz and Prof. Charles Ragin for very helpful feedbacks on various parts of the project. I also thank Prof. Paul Diehl and Prof. Robert Axelrod for useful comments on works related to different parts of the book. During the last year of the project I received much encouragement from Prof. Lotfi Zadeh whose excitement for the project provided much needed moral support and intellectual energy. Many friends, colleagues, students (especially my graduate students in the course on Fuzzy Logic Methods in the department of political science at the University of Florida), and family members provided valuable comments and encouragements without which this book might have not been finished. The book is much better for all this help and feedback.

I am also very grateful to Prof. Janusz Kacprzyk, editor-in-chief of the series on Studies in Fuzziness and Soft Computing at Springer, for his encouragement and smooth and speedy evaluation and processing of the manuscript which contributed to make the last stages of writing the book an enjoyable venture.

Above all, I am especially grateful to Khadidja Arfi, my wife, who never doubted the importance of the project and never took rest in encouraging me to pursue it until complete fruition. I dedicate this book to her. Thank you for your love and support Khadidja.

Acknowledgements

Contents

Chapter 1
Linguistic Fuzzy-Logic and Computing with Words

1 Setting the Ground

Data collecting, mining, and analysis are crucial aspects of research in social sciences. To this end a multitude of methods has been developed in various branches of social sciences. Yet the problem of vagueness of data expressed in natural languages is a persistent challenge that does not seem to be lending itself to a final resolution. Many years ago Charles S. Pierce stated that "vagueness is no more to be done away with in the world of logic than friction in mechanics." Likewise, Bertrand Russell retorted, "everything is vague to a degree you do not realize till you have tried to make it precise." Not surprisingly, vagueness and equivocation are constitutive features of any human system of communication. Human beings cannot live and communicate without using natural languages, yet the latter are inherently vague. Addressing the problem of linguistic vagueness is no easy task for social sciences. Indeed, most social sciences variables are often difficult to precisely operationalize, a problem that confronts equally those inclined to use either qualitative or quantitative methods of analysis. Constantly seeking more conceptual and operational rigor, precision and crispness is hence commonly believed to be a golden rule to the production of high quality research work in social sciences. Yet, the meanings of words are inherently, as literary critics and others are quick to emphasize, imprecise, vague, and fuzzy. Relying on rigorous quantitative methods does not solve this problem. Analysts might often fall into the problem of overdoing it, that is, by seeking too much precision and crispness in their methods they would end up sacrificing replicable connections with the phenomenon they are trying to explain. Moreover, the variables and approaches of these quantitative methods cannot ultimately escape a process of interpretation, which in the end more or less recreates the very condition of imprecision and fuzziness that conventional quantitative methods usually seek to avoid. It has thus become conventional wisdom in social sciences to seek better tools in addressing the problem of language-related vagueness in data analysis.

Contra conventional wisdom, this book raises the question: What if instead of seeking to get rid of language-caused vagueness we make it into an essential requirement for analyzing social science phenomena, both theoretically and empirically? This book thus develops an approach – termed as linguistic fuzzy-logic approach or LFLA – that addresses the problem of language-related

B. Arfi: Linguistic Fuzzy Logic Methods in Social Sciences, STUDFUZZ 253, pp. 1–17.
springerlink.com

vagueness in analyzing social sciences phenomena. Exploring the relatively new field of fuzzy-logic as applied to social and political phenomena takes us into a yet to be explored rich territory of research questions, theoretical arguments, and policy insights. If the engineering world has been fundamentally reshaped after Lotfi Zadeh (1999) launched his pioneering research program on fuzzy-sets theory in the mid-1960s, vigorously applying similar methods in social and political research undoubtedly promises to have a far reaching impact on our understanding and explanation of many social and political phenomena. The approach is fuzzy because it is built on fuzzy logic, instead of Aristotelian or Boolean logic. It is linguistic because it uses words as computational variables, not numbers.

The book hence proposes a new approach based on linguistic-fuzzy logic and the notion of computing with words to the analysis of decision making processes, strategic interaction, the study of causality, and data analysis. The book consists of theoretical discussions and developments of these new methods as well as their applications to various substantive issues of interest to international relations scholars, political scientists, and social scientists in general.

2 Fuzzy-Sets Theoretic Approach in Social Sciences

The notion of a fuzzy set, first introduced by Zadeh in 1965, is based on the idea that an object more or less corresponds to a category.[1] The concept of fuzzy set focuses on the vagueness that is intrinsic to natural language (e.g., in description such as "very smart" person, "widely used" terms) and that sets are fuzzy. For example, the sets of developing countries is fuzzy in the sense that it is both theoretically and empirically impossible to unambiguously and sharply define what a developing country is and where the boundary between the sets of developing and developed countries lies. This in fact is true for all states attributes such as democratic, democratizing, failing, consolidating, etc. If one were to use conventional ways to describe the set of democratic states, one would say that state A is either democratic or not – it is either a member or not a member of the set of democracies. As most students of democratization will testify though, this is a very un-empirical statement. In reality, states are always more or less democratizing. We would hence more accurately say that state A is more or less in the set of democratic states. The membership value is not necessarily 1 (i.e., "in") or 0 (i.e., "out'), as conventional wisdom based on crisp sets would have it. Rather, the membership value is somewhere between 0 and 1, with 0 corresponding to completely non-democratic and 1 corresponding to fully democratic. Dealing with this notion of gradual membership in a fuzzy set has resulted in a rich industry of concepts, theorems, algebras and the like.

Attaching a degree of membership to the elements in the fuzzy set means that the extent to which an element belongs to the fuzzy set is a number of the continuous interval [0,1], rather than on of the Boolean pair {0 (out) , 1 (in)}. A classical (crisp) set A is defined as a collection of elements $x \in$ A, the

[1] For and introductory text on Fuzzy Sets Theory and Applications see, for example: Smithson (1987); Klir and Yuan (1995); Smithson and Verkuilen (2006).

characteristic function *c(x)* of which can have one of two values, 0 or 1. A value of 1 means that element *x* is in the set A whereas a value of 0 means that it is not. The characteristic function of a fuzzy set allows *various degrees of membership* for the elements of a given set – the elements are included in the set to a certain extent or degree with some elements "more" included and others "less" included. Formally, a fuzzy set A (drawn from a larger set of elements, X) is characterized by a membership function μ(.) given by $\mu: X \rightarrow [0,1]$, and $0 \leq \mu(x) \leq 1$. $\mu(x)$ represents the degree to which *x* belongs to A. Compared with a crisp subset of X, the range of the membership function of a fuzzy set is a continuous mapping over the interval [0,1], rather than the two-element set {0,1}.

For example, in a fuzzy Prisoner's Dilemma game, the strategy of cooperation can be divided into a number of fuzzy sets (subcategories) described as *full cooperation*, *strong cooperation*, *moderate cooperation*, *weak cooperation*, and *no-cooperation*, with the corresponding membership functions given respectively by μ_F^C, μ_S^C, μ_M^C, μ_W^C, and μ_N^C. These are continuous functions on the interval [0,1], which means the players are more or less fully cooperating, more or less moderately cooperating, etc. Probability and statistics are insufficient to represent this sort of linguistically represented vagueness.[2] Fuzzy set theory is a tool that allows us to mathematically incorporate such an inherent vagueness in theories and methodologies of social sciences. More generally, as put by Ragin (2000:331), "paradoxically, fuzzy sets highlight the imprecision of social scientific concepts and demand that they be sharpened and clarified. The greater the correspondence between the scores indexing membership in a fuzzy set, on the one hand, and the meaning of the concept that parallels the set, on the other, the more useful the fuzzy set ... to construct a fuzzy set and assign fuzzy membership scores, researchers must say what their concepts mean, and they must be explicit about membership criteria."

In a ground breaking piece, Cioffi-Revilla sought to systematically introduce the ideas and concepts of fuzzy sets theory to the IR discipline in 1981. His goal was to show how "the ambiguity and uncertainty which are inherent in many historical alliances, decisions, and perceptions of international relations" are not the result of "random factors, unreliable quantitative data, and inaccurate measurement," (Cioffi-Revilla, 1981:129). He sought thus to highlight the potential for fuzzy sets theory to calculate and analyze fuzzy phenomena in IR for purposes of modeling and theory building. Although this piece did not *per se* go far in actually formulating fuzzy-sets theories of IR, it should be credited for alerting the IR readership to the potential power of fuzzy sets in addressing the inherent fuzziness of many IR phenomena and concepts. Not only did Cioffi-Revilla clearly present many of the ideas and tools of fuzzy sets as a succinct introduction for the uninitiated reader, he also showed through a variety of IR examples the usefulness, strength and potential of the approach. As he put it (1981:157-58), "the fuzzy approach can provide the framework within which to

[2] There are many new and very fruitful cross-marriages such as fuzzy probabilities, statistical methods for determining the membership functions of fuzzy sets, possibility theories, and so on.

develop truly general (although fuzzy) theories, from which current theories can be derived as special cases concerning well-defined classes in the universe of IR." This optimistic prognosis has not become a reality, unfortunately. Yet, a number of works have appeared since then which do indeed, if only partially so, confirm Cioffi-Revilla's expectations.

Most interestingly, G. S. Sanjian published a series of articles (1988; 1991; 1992) in which he formulates fuzzy-sets models of decision making in IR. His first piece (1988) presented a model of decision-making process of US arms trade using fuzzy set theory. The article (1988:1018) developed "a fuzzy set model of the process through which an arms exporting country selects a transfer for a prospective importer." It then presented "a method for deriving the exporter's optimal strategy in any arms trade setting." Finally, the piece provided an empirical test of the model "by comparing its ability to predict the policy choices of the United States for Third World importers with the predictive capabilities of two expected-utility models." The fuzzy sets model was able to predict successfully 91 percent of the cases tested. In addition, the model shows that "the politically advantageous strategy is always a function of (1) the relative importance of the countries in the import region of the exporter; and (2) the implication of the exporters' strategic options for its relations with those countries," (1988: 1043). Sanjian concluded that the fuzzy-set model is a good representation of the export patterns of a hegemon. In this respect, the fuzzy-set model performed as well as an expected utility model.

Sanjian elaborated this approach in 1991 to examine the process of arms transfers by the Great Powers. More specifically, he used fuzzy-sets theory to model the decision-making processes of hegemonic, industrial, and restrictive exporters of arms. Sanjian's fuzzy-set model outperformed rival autoregressive, minimum information, and expected-utility models with an overall success rate of 87 percent in predicting the exporting strategies of the United States, France, and West Germany to the Third World in the period 1959-1976. Sanjian (1991:192) concluded that "much of the hegemonic, industrial, and restrictive decision making can be explained with a relatively parsimonious nontraditional formal model." He further developed his fuzzy sets approach to examine the collective decision making processes within NATO, specifically considering the 1989 debates on whether to modernize the alliance's short-range nuclear missiles or negotiate an agreement for force reduction with the Warsaw Treaty Organization. After summarizing the fuzzy-sets approach to collective decisions under multiple criteria, Sanjian applied it to the case in point, concluding that the model successfully predicted NATO's adoption of Germany's position on the issue.

The idea of using fuzzy sets in social sciences was given an important boost when Charles Ragin published his 2000 book on "Fuzzy-Set Social Science." Much like Cioffi-Revilla and Sanjian, Ragin sought to explore the utility of fuzzy-sets theory as a way to bridge the divide between qualitative and quantitative methods, arguing in this regard that fuzzy sets allows a far richer exchange between ideas and evidence in social research than what the existing methods offer. Ragin uses fuzzy-sets theory to refute conventional "homogenizing assumptions" of variable-oriented approaches to empirical cases and causality and instead advances an agenda based on diversity-oriented research strategies. Most importantly, Ragin concisely shows how fuzzy-sets theory allows a natural fit between evolving theoretical concepts and in-depth knowledge gained through case studying. Ragin's book is very rich in insights and discussions about a variety of issues that pertain to the discovery process through which fuzzy-set social science evolves. Moreover, he devotes a large portion of the book to exploring the issues of necessary, sufficient, and complex causality from a fuzzy-set perspective. Not only does he show how to "fuzzify" our conventional treatments of causality. He also uses real examples from international political economy (such as protests against IMF policies) to illustrate how a fuzzy-sets approach to causality would perform in empirical testing.

Although the various disciplines of social sciences have yet to fully recognize the potential of fuzzy-sets methods to theory formulation and testing, these works clearly make a good case that fuzzy sets methods are at least as good as many other methods, if not better as this author believes. Not surprisingly, there are problems with the above works as they apply fuzzy-sets theory to social sciences. One key shortcoming is the arbitrariness that the notion of membership function suffers from, especially when applied to empirical cases. Although this is not specific to social sciences and phenomena, it becomes the more important because all the above cited works prescribe a core role to membership functions.

Membership functions are generally speaking at the core of many fuzzy-sets theories. This prompted many skeptics about the import of fuzzy sets for social sciences to argue that determining the membership function is more or less both theoretically and empirically arbitrary due to two major problems which emerge when using fuzzy sets to theorize about concepts by relying on the notion of membership functions. First, there is no generally agreed upon method to derive or calculate membership functions, either from first principles or empirical data. Second, uncertainty and vagueness can also occur in the membership functions themselves. The degrees of membership to a category can also become fuzzy sets themselves, leading to the concept of type-2 fuzzy numbers whose membership functions are fuzzy sets. The transition from ordinary crisp sets to type-1 fuzzy sets is done when we cannot determine the membership of an element in a set as either 0 or 1, that is, when the membership becomes any number between 0 and 1. Similarly, when the circumstances are so imprecise and fuzzy that we have trouble determining the membership degree as a crisp number in the interval [0,1], we can move to type-2 fuzzy sets, that is, we turn the membership functions into fuzzy sets themselves. The problem is that this process is in fact *ad infinitum*, resulting potentially in type-n (and, in principle, type-∞) fuzzy sets. This makes the manipulation of fuzzy sets and numbers opaque and reduces the tractability of these tools in dealing with empirical problems.

These problems notwithstanding, much improvement has occurred in the field of fuzzy-set theory to make the approach much more attractive in social sciences. In this book, I present one approach wherein the membership function does not play an important role, both theoretically and in empirical data analysis. This approach based on linguistic fuzzy logic draws on the idea that it is possible to avoid dealing with the problems due to the arbitrariness of membership functions by not using them in the first place. This is done by going back to the very first notion of fuzzy sets theory which is to preserve the "fuzziness" of natural language using variables the values of which are "words."

3 Why Fuzzy Logic?

Every theory and every empirical analysis of social phenomena is underpinned in a given logic of reasoning for judging coherence, consistency and the possibility of falsification. Much of what counts as social science today is based on a Boolean logic with two truth values, 0 and 1. The task set out for this book is how to go beyond this assumption by showing that linguistic fuzzy logic can be very useful in analyzing empirical data, studying causation, and strategic interaction in social science theories. Doing so would enable us to express and analyze the role of language-related irreducible uncertainties and vagueness assuming various manifestations in validating our theories about the human world. As a result we would thereby be able to capture more accurately and faithfully the intricacies of human reasoning, cognition, and communication as they inhere in human language.

The assumption of a Boolean logical framework runs counter to many of our observations and conceptualizations about social/political phenomena. For example, we know that it is both theoretically and empirically impossible to unequivocally and precisely define what a developing state is and where the boundary between the sets of developing and developed states lies. However, we often ignore this problem simply by assuming Boolean logic as the underlying logic of our conceptual apparatus and analysis – that is, by saying that a state is either developed or not developed. One can argue that we may instead characterize the state as developing with a higher or lesser level of development. However, this does not alleviate all forms of equivocation since to carry out the empirical analysis, we definitely need to classify the set of so-characterized developing states into categories that do not admit overlapping cases; that is, boundaries between categories would need to be sharply delineated. The process of classifying involves verifying that the cases can be unambiguously assigned to the appropriate categories. Thus, should we be unable to categorize a case, we would not include it in the analysis. Put differently, a case can belong or not belong to any single category – in-between positions described as belonging more or less to two or more different categories are not allowed to exist in Boolean logic thinking. One way to address this problem of overlapping categories and fuzzy boundaries is to reconsider the assumption of Boolean logic underlying much of what counts today as social sciences. Doing so has a number of benefits.

First, a non-Boolean approach would radically redefine our conception of what is a consistent and coherent theory. There is widespread belief that we can ascertain the consistency of our analyses and formal theories only through an exploration of their underlying logic. This belief is rooted in one unspoken assumption – that is, to take the principle of Boolean contradiction as a fundamental pillar in defining logical consistency. The interesting exchange between Stephen Walt and his critics on the criteria with which we should judge formal theories in international security studies is illustrative (Brown et al. 2000). All participants agree that a good theory should be logically consistent. Formal and mathematical modeling is praised by its advocates precisely because it is believed to help guard against inconsistencies due to logical failure of the theory or resulting from ambiguities inherent in natural language. As put by Zagare, "There can be no compromise here. Without a logically consistent theoretical structure to explain them, empirical observations are impossible to evaluate; without a logically consistent theoretical structure to constrain them, original and creative theories are of limited utility; and without a logically consistent argument to support them, even entirely laudable conclusions . . . lose much of their intellectual force" (Brown et al., 2000:103).

A resolution of logical inconsistencies through the exploration of the logic of a theory is thus a central part of the scientific endeavor. These notions of logical coherence and consistency are assumed to be Boolean (true/ false) in nature. This is well reflected in the rhetoric of "paradox." To highlight the value added of their work, many political/social scientists phrase their research questions as a paradox. However, what goes unnoticed in this deployment of rhetorical power is that this paradox is only a paradox given the posited underlying logic. For a different logic—and there are so many other logics besides Boolean on-off logic—this paradox might not be a paradox at all but rather a consequence of the axioms of the logic. For example, paraconsistent logic is built around precisely negating this principle (Heyting 1971; Priest, Routley, and Norman 1989).[3] A theory built on the premise of a paraconsistent logic, which inherently takes contradiction as an axiom, will undoubtedly have no room for such an argumentation procedure! Likewise, starting from so-called intuitionistic logic, which is a classical logic without the Aristotelian law of excluded middle: ($A \vee \neg A$), would logically validate Boolean logic–type inconsistencies in a theory. This hence raises a serious question on the conventional wisdom behind the requirement of logical consistency—it becomes contextualized (i.e., dependent on the underlying logic).

Conventional wisdom has it that "an inconsistent theory creates a false picture of the world. Inconsistent theories are also more difficult to test because it is harder to know if the available evidence supports the theory" (Brown et al., 2000: 8). I suggest that, instead of taking it for granted that "logical inconsistencies" are the first culprits to be eliminated from a theory as a way of improving its empirical validation, we need first to realize that this "logical inconsistency" is contextual (i.e., there is an inconsistency only within the purview of the Boolean logic that is being assumed). To avoid any misunderstanding here, let me be clear that I am not calling for sloppiness and anything-goes attitude. To the contrary, I am suggesting that we need to put our research endeavor on firmer logical grounds by being precisely clear on the logic that underpins our theorizing and empirical testing works. Being clear at the forefront about the limitations of one's theory and

[3] The liar paradox "This sentence is not true" is illustrative. We have two options – either the sentence is true or it is not. (1) If we suppose it to be true, then what it says (i.e., this sentence is not true) is the case. Therefore, the sentence is not true. (2) If we suppose that it is not true, which is what it says, then the sentence is true. In either case (1) or (2), it is both true and not true.

method is highly praised in social science. However, we often stop short of fulfilling this commitment to the full extent, right down to the logic underlying our theories and methods. This is problematic because understanding the inner working of the underlying logic and how it is axiomatically constructed opens up the possibility for considering other types of underlying logics of relationships among variables – and there is no shortage of other logics (e.g., fuzzy logic, modal logic, intuitionistic logic, paraconsistent logic, quantum logic, etc.). Boolean logic still largely underpins the overwhelming majority of works in social sciences for the most part due to genealogical reasons and resistance due to path dependency. Yet, there are no ontologically, epistemologically, theoretically, empirically, or methodologically insurmountable impediments or raison d'être why we cannot explore the impact of the vast realm of other logics on the theorization, design, and execution of social science inquiry. As the burgeoning literature on the fuzzy-set theoretic approach to social, economic, and political inquiry testifies, we have much to gain from doing so.[4] This book seeks to contribute to this burgeoning literature.

Second, espousing fuzzy logic instead of Boolean logic has a far-reaching impact on how we think of and analyze causality in social sciences. Generally speaking, we depend upon causation all the time to explain what happens to us. We use causality statements to draw realistic predictions about what might happen as well as to direct what might happen in the future. In short, we are in constant searches for causal explanation and behavior. A widespread motto is, to put it formally, "X caused Y" or "Y occurred because of X." Not surprisingly then, studying and explaining as well as modeling causality is at the core of much of what counts as social sciences today. Although other types of analysis (such as constitutive) are important aspects of our inquiry about the social and political world, causal analysis absorbs much of our collective effort. Yet, the tools that have been designed for this purpose remain rather limited in social sciences.

For the most part scholars use statistical means to analyze data and explore causality. However, as succinctly put by Braumoeller (2003:209), "theories that posit complex causation, or multiple causal paths, pervade the study of politics but have yet to find accurate statistical expression ... To date, however, no one has made a concerted effort to describe how the empirical implications of theoretical models that posit causal complexity could be captured by statistical methods." Belittling neither the achievements nor the gap in statistical methodology that Braumoeller highlights, there is a need for other types of tools – non-statistical approaches – that could be used to address issues that statistical analyses cannot reach such as conceptual vagueness. As argued by Charles Ragin in making his case for using fuzzy set theory in social science analysis (2000:5), this would help us relinquish many of the homogenizing assumptions that underpin much of quantitative analysis such as viewing populations, cases, and causality. This would hence undoubtedly strengthen the link between empirical data and theory.

Probabilistic analysis of causality has the source of much insight. The approach draws on the theory of probability, hence defining causality in terms of probabilistic relationships. In fact, probability theory is the most used

[4] See the website http://www.compasss.org for some current literature in this respect.

mathematical language in dealing with causality in most of what counts as social sciences today. This is not surprising since uncertainty is ubiquitous in social settings and probability theory allows us one powerful way with which to express this uncertainty, such as in terms of likelihood estimates for connections between antecedents and consequents of causal relationships. Moreover, probability theory allows us to deal with situations where there are exceptions to the rules of deterministic logic. The basic idea of this type of causality is that a cause should increase the probability of its effect. This can be formally put as: A is a cause of B if and only if B is more likely in the presence of A than in the absence of A, *ceteris paribus*, that is, in terms of the conditional probability: $P(B|A) > P(B|\sim A)$. Scholars also use formal and mathematical modeling to explore causality. Although not without its own problems (such as stylization-induced issues) formal modeling is particularly suited for the study of the logic of causality since it is rigorously designed and its conclusions are clearly and unambiguously drawn from well stated assumptions and limitations.

Whether it is statistical/probabilistic or formal/mathematical modeling in nature the analysis of causality assumes a given logic and an underlying algebraic structure of this logic. In most studies of causality in social sciences this logic is taken to be the Boolean logic with its associated two-valued (false, true) algebraic structure. The focus on the underlying logic is important because only through an exploration of the underlying logic can we ascertain the consistency and completeness of our analysis of causality, whether we are dealing with necessary and sufficient conditions or with multiple/complex causality. Understanding the inner working of the underlying logic and how it is axiomatically constructed opens up the possibility for considering other types of underlying logics of causal relationships.

As an illustration of the great potential of considering different underlying logics in studying causal arguments, let us briefly see how fuzzy logic might reshape the seemingly never-ending debates in the literature on democratic peace. In a very innovative work, Zinnes (2004) uses propositional calculus (based on Boolean logic) to explore the logic underpinning the various arguments of the democratic peace. As she puts it, "Just as we moved from initial democratic-peace statistical results to explanations for them and then to critiques of those explanations, we now need to take the next step and expand and revise the logics to better incorporate such observations as the war initiation behavior and war victory rate by democracies" (p. 453). Yet, the "logics" that Zinnes is referring to are all underpinned by Boolean logic. The fact that Zinnes was able to logically validate the normative and institutional arguments of the democratic peace using Boolean propositional calculus raises the following question: What if we were to use a different foundational logic, such as fuzzy logic? Would this help us make more sense of the democratic arguments and the diverse and often mutually contradicting, or at least, ambiguous, results of its empirical studies? A fuzzy-logic approach to the problem would, for instance, rephrase the basic propositions that Zinnes starts from very differently with a strong potential of not only reconfirming Zinnes's conclusions but also going beyond them, thereby raising new questions that are outside the purview of the Boolean world. Analyzing the democratic peace through a fuzzy-logic approach will definitely lead to a different propositional setup and broader conclusions.

Mentioning a few examples suffices to make the point (LFL counterparts are in italics). When Zinnes (2004) writes, "If state is a democracy, then all decisions involve the participation of the population and its representative institutions," or "If state is a non-democracy, then all decisions are made by a small group of elite leaders," the fuzzy-logic counterpart will be "*If state is a more or less democracy, then all decisions more or less involve the participation of the population and its representative institutions.*" Likewise, when Zinnes writes, "If a state uses bargaining to settle internal societal conflicts and its security is not threatened, then it will use bargaining to settle internation conflicts," we would have "*If a state more or less uses bargaining to settle internal societal conflicts and its security is more or less threatened, then it will more or less use bargaining to settle internation conflicts.*" Similarly, "If states X and Y are in conflict and bargaining is used to settle internation conflicts, then states X and Y do not use force to settle internation conflicts" would turn into "*If states X and Y are more or less in conflict and bargaining is more or less used to settle internation conflicts, then states X and Y more or less use force to settle internation conflicts.*" We can see that this produces much vagueness in the various propositions. Thus, in fuzzy logic, we can conclude that two states can go to war or not go to war—just like in Boolean logic. We also can conclude that two states can more or less go to war—that is, the two states are in a no-peace situation but not in a fully-fledged state of war. Although it is premature to advance any concrete conclusions about a fuzzy-logic approach to the democratic peace, it nonetheless opens up a new way to addressing the still-unresolved issue (see Chapter VI for further considerations on the democratic peace argument). In sum, explicating the logic underpinning the analysis opens up the horizon for new questions, new methods of thinking about and investigating social and political phenomena, and new results. This would help us relinquish many of the homogenizing assumptions that underpin much of quantitative analysis such as populations, cases, and causality (as pointed out by Ragin 2000: 5).

Third, ironically, what makes Boolean logic a simple, rigorous basis of analysis is also its weakest point – dichotomization of the truth values of the antecedent and consequent variables. Boolean logic forces us to think only in terms of true and false statements. Yet, as the wealth of statistical and qualitative studies show, vagueness is inescapable in all our theoretical and empirical analyses of social and political phenomena. Statements that such and such variable causes more or less such and such outcome are pervasive in all social sciences. What is important to note here is that this vagueness is not just a type of epistemic uncertainty that might be resolved with better tools of analysis. Rather, linguistic vagueness goes deep down into the logic underpinning our analyses. This vagueness eludes Boolean dichotomization and hence drops out in the stylization process of formal theories. A key idea in fuzzy logics is to inscribe this vagueness in the logic of our theories, positing that the truth value of a proposition can assume other values than just "true" or "false." Thus, it is incorrect to think that to know the world, it is sufficient to know the "truths"; we also need to know the "falsities," as well as the degrees of both. This is very different from Boolean logic, which upholds that: *If a proposition p is not true, then p is false. Should we thus know that p is not true, there is no need to inquire whether p is false? it is taken for granted to be false.* There are many situations in daily life when this logic is simply not applicable. A simple example from ordinary language usage is when one answers by stating, more or less yes and more or less no. This is beyond Boolean logic!

Fourth, in Boolean logic, it does not matter much whether one uses either {0, 1} or {True, False} to express the truth values of the logical propositions. In fuzzy logic, the choice matters. Indeed, we can have a multivalued fuzzy logic, the truth values of which are expressed using numerical values belonging to the set [0, 1]. However, this version of fuzzy logic still relies on the concept of membership function (i.e., the degree of belonging to a set) and runs into many problems, especially in using membership functions as a tool to analyze the sociopolitical world (Ragin 2000). Linguistic fuzzy logic is one way to go beyond the problems and difficult-to-resolve issues that are generated by these numerical membership functions.

4 Why Linguistic Fuzzy Logic?

A key insight behind a linguistic fuzzy logic approach is that we can we use natural language to express a logic in which the truth values of propositions are expressed in natural language terms such as true, very true, less true, less false, very false, and false, etc., instead of a numerical scale. The Linguistic fuzzy logic approach that I propose in this book is based on a basic set populated with symbols (more precisely, words and qualitative modifiers or hedges). Linguistic fuzzy logic allows us to study graduality in social phenomena using natural language.

This book is in some sense a continuation of, and provides an improvement to, the method of published works on fuzzy logic methods in social sciences. However, these works almost exclusively use a *numerical* format for fuzzy sets theory, as illustrated by Charles Ragin's 2000 book on "Fuzzy-Set Social Science." This book goes beyond Ragin's work and similar others to introduce new *linguistic* fuzzy logic methods. Although this work is still anchored in fuzzy-sets theory there are two main features that make it stand well apart from similar works published in social sciences. First, all published works are interested in using numerical fuzzy-sets theoretic tools as the basis for new methodologies for empirical analysis. My approach seeks to analyze social science phenomena using tools developed based on linguistic fuzzy logic. Second, all published works are based on the use of what is known as membership function or degree of belongingness to sets, a number between zero and one. My approach computes with words all the way down and does not use membership functions (Zadeh, 1999). I instead consider the degrees of belongingness to a category as a linguistic variable termed as a nuance value. I also introduce a second type of fuzziness at a deeper level than what extant works consider. The membership degrees are also fuzzy sets, or more exactly, fuzzy linguistic values, termed as truth values of the nuance values. We would say, for example, that country A is democratic with a very high nuance; that is, it is very highly democratic. We also say that nuanced democracy is, for example, true to a low degree. This degree of truth would then measure our confidence level in the nuance level of democracy.

The approach contributes to shed light on lingering disputes on tradeoffs between rigor in reasoning, precision in conceptualization, correspondence with empirical reality, and logical consistency. As all social scientists very well know there is a constant vagueness that shapes all theories and analyses of social sciences. Most of what counts as mainstream social science today assumes that this vagueness is an epistemic problem that can be reduced using increasingly more sophisticated statistical and otherwise tools in formulating our theories and analyzing empirical data. The linguistic fuzzy logic approach posits that vagueness is essentially not epistemic. Vagueness is inherent to the constitutive and causal logic underlying the phenomena themselves. From this perspective, using Boolean two-valued logic as a means of systematically studying the consistency and coherence of theories might very well distort the very logic of the phenomenon under study. Our minds are language-shaped and as such the logic of the workings of our minds should be language shaped. We do not live in a binary world of 0s and 1s. We live in a world the logic of which is best understood using linguistic expressions, not numbers.

Assuming that the logic underlying social sciences is of a linguistic fuzzy type has important implications for research in social sciences. First, much ink and talk have been spent on the qualitative-quantitative divide in social sciences, thereby contributing to enrich our approaches and knowledge about the world. However, much of this debate does not go far enough from the perspective of this book. Although logic informs all reasoning, whether the latter is based on quantitative inquiry or qualitative work, many researchers are not fully aware of the implications and limitations of the logic underlying their inquiry and how this logic informs the epistemology which supports their empirical work. Linguistic fuzzy logic approach is one important step toward bridging the gap, or at least offering a new perspective, on the ontological side of the debate which has for the most part been dominated by methodological issues.

That this is the case has much to do with the fact that social science theory is a discourse and just like any other discourse it is engulfed in larger theoretical and social discourses. As stated by Layder (1985:255), "knowledge of this world is impossible without the use of conceptual instruments, which, more often than not, derive from, or are connected with, wider theoretical parameters or discourses." This should hence warn us against "making closed or dogmatic claims about the irreducible ontological features which societies possess and hence the terms in which such features can be known," (Layder, 1985:255). In other words, theoretical discourses play a role in determining social ontologies. Moreover, "epistemological and ontological levels [of theories] are tightly bound together through the mediating lens of the linguistic/conceptual structure of the discourse" (1985:261). Social scientists cannot fail to recognize "the constructional role of discursive elements in relation to ontology, (and thus the discursive relativity of social ontologies and ontological features)" and should give up "insisting on the search for ontological features independent of discursive knowledge" (Layder, 1985:273). No discourse – including discourse about ontology, epistemology or methodology – is outside the more or less strong influence, if not complete determination, of preexisting linguistic and conceptual as well as other discursive contexts. This insight forces us to rethink the delineation of the discourse and categorization of ontology from epistemology (and methodology, etc…). The very categories of social science jargon such as ontology, epistemology, methodology, etc…, are always deployed within discourses that seek to be consistent and logical. More often than not we seek to sharply define our concepts and categories as much as we can. We also seek as much as possible to avoid contradictions, especially, self-contradictions. Yet we know that natural language is inescapably vague and imprecise, despite incessantly brave efforts by people to sharpen the semantics. Thus, we like to think that categories such as ontology and epistemology can be defined in such a way that the two do not overlap. One of the key features of linguistic fuzzy logic is the logical possibility for non-binary truth values. That is: linguistic fuzzy logic allows us to assert that a statement about theory (or anything else) is *more or less truly* ontological, *more or less truly* epistemological, *more or less truly* methodological, etc…. In linguistic fuzzy-logic terms the categories of ontology, epistemology, and methodology are always *more or less* overlapping! Therefore, if instead of implicitly relying on a binary Boolean logic in thinking about ontology, epistemology, and methodology, we were to use linguistic fuzzy-logic as a conceptual framework, the conventional position about delineating ontology from epistemology and methodology would fall asunder, or at least be fundamentally revised.

Second, both qualitative and quantitative research engage in data reduction, with the former doing the reduction by using words, categories, and themes and the latter doing the reduction by using numerical and often statistical tools. The analysis of data (after reduction) is done in qualitative research through categorizing and comparing and in quantitative methods through statistical inference and estimation of likelihood and the like. Of course, these differences are matters of degree. However, generally speaking, quantitative research is believed by its practitioners to generate reliable population based and generalizable data and is hence well suited for studying cause-and-effect relationships. The fact that qualitative data typically involves words and quantitative data involves numbers prompts many researchers to believe that the latter is better and more scientific than the former. The linguistic fuzzy logic approach proposed in this book is one way to create a bridge between the qualitative and quantitative sides of this debate. The approach integrates some of the best features of both sides. The linguistic fuzzy-logic approach is both rigorous and incorporates vagueness at its very heart. It proposes mathematically rigorous algorithms to reduce and analyze data which is expressed in a natural language, hence preserving the vague meanings of linguistically defined variables.

Third, a key contribution of linguistic fuzzy logic approach to the study of causation and other processes in social sciences is to emphasize the need not only of measuring the linguistic values of both independent and dependent variables, a call that regular fuzzy-set theoretic approach already makes. The approach of the book also shows the crucial role of a dimension of both independent and dependent variables which by and large has been completely ignored in measuring and analyzing empirical data, namely, the truth value of the linguistic values of the variables. Hence, linguistic fuzzy-logic analysis is more demanding in terms of empirical data than Ragin's fuzzy-sets theoretic approach because it requires not only a fuzzification of the values of categorical variables but also an estimation of our confidence in the evaluation of these fuzzified values. As shown in the various illustrations considered in this book, different values of the truth levels do shape the final conclusions of the analysis. A comprehensive and systematic analysis of data and causation and other processes cannot ignore this crucial aspect of the analysis. Overall, this linguistic recasting of fuzzy logic approach has a number of far-reaching consequences for social science analysis. First, linguistic fuzzy-logic approach reaffirms the importance of fuzzy-set theoretic analysis as developed by Ragin and others. However, going much beyond these works, the approach of this book shows how to avoid resorting to mixing fuzzy-set theory and statistical analysis in order to achieve a comprehensive analysis of causality (as Ragin, for example, does in his book). The concept of linguistic truth level, which naturally emerges in linguistic fuzzy logic, makes the book's approach self-contained. In a nutshell: Extant fuzzy-set approaches deal only with the nuance dimension of fuzzy logic, whereas linguistic fuzzy-logic approach deals with the nuance as well as the truth levels of these nuance levels, on an equal footing.

Fourth, linguistic fuzzy-logic approach enriches our very notion of causation itself. Conventional practice and conceptualization discuss causation in terms of necessity, sufficiency, and combinations thereof. Boolean approach leads to the common view that necessary and sufficient conditions are dichotomous in nature. Something is either necessary or it is not, something is either sufficient or it is not (Goertz and Starr, 2003:3). According to this view, our commonsensical statements that this or that condition is "virtually necessary or sufficient" or "almost always necessary or sufficient" do not make much sense in social science analysis based on Boolean logic. This book shows that such notions are too restrictive given the richness of natural language which is constitutive of social reality as we construct it, experience it, and analyze it. Instead of talking about necessity and sufficiency, as most of what counts as social science practice does, linguistic fuzzy-logic approach offers to us notions of fuzzy causation. Hence, for example, instead of talking about crisp necessity we should speak of fuzzy necessity, which is a concatenation of necessity and contingency, a notion quite different from probabilistic necessity which is essentially still based on crisp notion of necessity. In linguistic fuzzy-logic approach, crisp necessity and crisp contingency become limiting cases of fuzzy necessity.

5 Organization and Plan of the Book

The book has several chapters, successively dealing with the process of fuzzy decision making, fuzzy strategic interaction and fuzzy game theory, fuzzy causality and fuzzy propositional logic, and fuzzy data analysis. Overall, this manuscript develops a new way of thinking about fuzzy-logic methodology in the study of social and political phenomena. Not only does it go beyond already published works based on fuzzy-sets theoretic methods, but it also presents new ideas as no other book in social sciences or elsewhere has ever done before. The book also makes a strong case through its scope and rigor for adopting linguistic fuzzy logic, thereby introducing the subject-area of "computing with words" into social sciences methods. The book specifically plows new grounds by presenting new kinds of game theory, and data analysis methodology. The book hence takes social science methodology well into yet to be charted territories of research and analysis.

Chapter 2 presents a number of elements of linguistic fuzzy logic needed in the subsequent chapters such as the notion of fuzzy sets, algebra of linguistic fuzzy variables and the methods of aggregating linguistic fuzzy information. Chapter 3 develops a linguistic fuzzy-set theoretic approach to study decision making. This chapter models how actors who are faced with situations fraught with vagueness make their decisions. The work considers situations where the actors engage in a process of evaluating the merits of different choice alternatives under multiple criteria – a process of Multiple Criteria Decision Making. A key element of the method is that it avoids an unnatural numericalization of knowledge that is expressed linguistically. How do decision makers aggregate the linguistic information that they have on vague criteria to reach a final decision? A conventional way to study the process of decision making under multiple criteria is by comparing the costs and benefits of the possible choices that decision makers face and then conclude that they would opt for the optimal choice. To this effect, practitioners and analysts do their best in achieving a very high degree of precision in estimating the various variables and criteria that are postulated to determine the final choice. This is no easy task because both independent and dependent variables are not easy to operationalize. The meanings of words are inherently imprecise, vague, and "fuzzy;" a problem that relying on numerical methods does not solve. Linguistic fuzzy-set theory allows us to address these issues. The linguistic fuzzy-set approach also contributes to a linguistic-fuzzy form of social choice theory. It allows the possibility of paying more attention to imprecise preference relations and presents aggregation tools that can help explore the possibility of social choices. In short, the linguistic fuzzy-set method can be applied to individual decision making as well as collective choice situations. The approach is illustrated through a running (hypothetical) example of a situation in which state leaders need to decide how to combine trust and power to make a choice on security alignment.

Chapters 4 and 5 develop a new game-theoretic approach based not on conventional Boolean two-valued logic but instead on linguistic fuzzy logic. The latter is characterized by two key features. First, the truth values of logical propositions span a set of linguistic terms such as true, very true, almost false, very false, and false. Second, the logic allows logical categories to overlap in contrast to Boolean logic, where the two possible logical categories, "true" and "false," are sharply distinct. Linguistic fuzzy logic allows us to study graduality in strategic interaction using natural language. A conventional game can become a linguistic fuzzy game by turning strategies into linguistic fuzzy strategies, payoffs into linguistic fuzzy payoffs, and the rules of reasoning and inferences into linguistic fuzzy reasoning, which operate according to linguistic fuzzy logic, not Boolean logic. This leads to the introduction of a new notion of fuzzy domination and Nash equilibrium which are based not on the usual greater than relation ordering but rather on a more general form of relation termed linguistic fuzzy relation. The linguistic vagueness that inherently exists in real strategic situations is not considered an epistemic issue but rather an ontological one and is incorporated in the very logic underpinning the reasoning process. One immediate result is that a linguistic fuzzy strategy can perform much better than a crisp strategy in dealing with multiple Nash equilibria. Linguistic fuzzy games can also be extended into N-person linguistic fuzzy games. In a linguistic fuzzy N-person game, each agent models others as linguistic fuzzy rational agents and tries to find a linguistic fuzzy Nash equilibrium that will achieve the highest linguistic fuzzy payoff. If the linguistic fuzzy relation is simplified into a crisp two-valued logic, the linguistic fuzzy game reduces to the conventional games. As such, conventional game theory can be viewed as a special case of the linguistic fuzzy game. I apply the new formalism to situations of PD game, Chicken game, Battle of the Sexes game, a trust game, and social game of cooperation.

Chapter 6 explores new ways to study complex causality in social sciences using a linguistic fuzzy-logic approach. This method allows us to formally and with mathematical rigor address a core problem – vagueness and lack of precision – in research design and theory testing in social sciences. Without belittling the tremendous amount of efforts that has gone into sharpening our theoretical concepts, operationalization, and empirical testing for both simple and complex causality for centuries, this chapter builds on Charles Ragin's ideas on fuzzy sets and offers a linguistic fuzzy logic framework where complex causality as well as ontologically constitutive relations can be "naturally" studied using words that are essentially vague, much like the phenomena that we seek to explain and understand. This is done by taking into account the possibilities that certain causal variables might be *more or less* of a sufficient type while others might be *more or less* of a necessary type, and while still others might be of both types to a lesser or greater degree of truth. The method is illustrated by examining the propositional-logical validity of the causal arguments made on the theory on the democratic peace. In brief, while the linguistic fuzzy analysis does not prove the democratic peace argument to be wrong, it adds much nuance to it. We do not have only options of War and Non-War. We also have many more possibilities of more or less war and no-war depending on whether the states are more or less democratic. This linguistic fuzzy logic conclusion provides, if partially, a logical explanation to the still ongoing debate on the empirical validity of the democratic peace argument. One can arguably say that: the problem seems neither to be one of empirical validation, nor one of theoretical explanation (although these are still somewhat debatable). Instead, from the perspective of this work, the problem seems to have much to do with the taken-for-granted Boolean logic approach which underpins the existing literature on the democratic peace. The approach of the chapter is also applied to a theoretical falsification of Skocpol's theory of revolution.

Chapter 7 presents a new approach to data analysis and the study of multi-path, multilevel causality in social sciences using linguistic fuzzy logic as a framework. The approach presented in this chapter differs from conventional data analysis and the study of causality on two fronts. First, all variables have two degrees of freedom (or levels of variation): a linguistic nuance value, which corresponds to what we conventionally refer to as interval or categorical value, and a linguistic truth value, which measures our confidence level in this nuance value. Second, combining this double fuzzification of variables with linguistic fuzzy logic the book proposes new tools of fuzzy data analysis and the study of fuzzy causality. More specifically, this new methodology is compared and contrasted with Ragin's (2000) fuzzy-set theoretic approach to data analysis and the study of causality and Braumoeller's (2003) Boolean approach to multi-path causation. These comparisons are illustrated through a re-examination of Ragin's study of IMF protest and Huth's (1996) theory of war initiation. The new method enriches our understanding and modeling of the very notion of causation itself.

The final chapter of the book concludes by highlighting the value-added of a linguistic fuzzy-logic social science. Because the book suggests new ways of thinking about and analyzing decision making, game theory and strategic interaction, data analysis, and theory falsification, both at the theoretical and empirical levels, it should appeal to a diverse pool of established scholars as well as to graduate students seeking to learn new methods of analysis. The approach of the book also helps improve the connection between the scholarly and policy worlds. Although there are many reasons why such a linkage does not always evolve in mutually beneficial and constructive ways, one impediment that seems to reinforce the lack of effective communication is that the media used by scholars in their research are often times very opaque to the practitioners. Conversely, scholars must often play the role of translators between the policy world and their theories about social and political reality. This process is, as we all know, fraught with many slippery turns. The linguistic formal method proposed in this book lends itself quite naturally to provide an effective means of communication between scholars and practitioners since the approach greatly facilitates the exchange and analysis of "raw" linguistically expressed information. Scholars and their theories and hypotheses can directly "speak" to the policy world and incorporate "raw elements" of the policy world in their theories and hypotheses.

Chapter 2
Elements of Linguistic Fuzzy-Logic and Framework

This chapter sets the ground for all subsequent ones by introducing key elements upon which the linguistic fuzzy-logic approach is built. Due to the novelty of linguistic fuzzy-logic in social sciences and the need for rigor in presenting the material, the chapter is somewhat demanding in its formal aspects. Yet the mathematical manipulations themselves do not involve any advanced knowledge of mathematics as such. Most importantly, the formal aspects are imperative for an understanding of key elements in subsequent chapters. For example, some readers might wonder why I include a discussion of formal logic. This will become clear since, as explained for example in chapters 4 and 5, I go back to the formal logic level of conventional game theory and then use that as a springboard for introducing the linguistic fuzzy-logic version of game theory. I also do likewise for introducing linguistic causal and data analysis and linguistic network analysis. In short, although this chapter might at a first look seem to be mathematically demanding, technically speaking it is not, except at the level of formal rigor. The chapter begins with a discussion of how one goes from Boolean logic to fuzzy logic analysis. It then presents a discussion of the essential notion of linguistic variable and related issues. The chapter ends with a discussion of how to aggregate linguistic information expressed using linguistic variables. I include illustrative examples throughout the chapter to make the material less dry as much as possible, without however losing the imperative of formal rigor.

1 From Boolean- to Fuzzy-Logic Based Analysis

Boolean logic is the logical framework of most of what counts as social sciences today. Boolean logic is an example of a system of formal logic. Generally speaking, a system of (formal) logic consists of a language equipped with a deductive system and/or a model-theoretic semantics.[1] The deductive system is meant to codify which inferences are correct for the given language, and the semantics is to capture the meanings termed as possible truth conditions for at least part of the language. As an illustrative example which explicates its reliance on Boolean logic consider Zinnes' (2004) work using propositional calculus to explore the logic underpinning the democratic peace. Zinnes seeks to ascertain

[1] Much of the mathematics on logic (theorems, definitions, and the like) that I use in this article is drawn from the following two books: Novak, Perfilieva, and Mockor (1999) and Hajek (1998).

B. Arfi: Linguistic Fuzzy Logic Methods in Social Sciences, STUDFUZZ 253, pp. 19–41.
springerlink.com

which arguments are sufficient to predict peace between democracies and war otherwise. This requires "identifying the key ideas in each argument, the atomic propositions, and the construction of the central premises of each theory by linking atomic propositions using the operations of *not*, *and*, *or*, and *implies*" (Zinnes, 2004:431).

Propositional Logic is the most rudimentary form of formal logic. At the most basic level, all knowledge is made up of propositions. In Boolean propositional logic propositions are either true or false. There are thus two truth values, "true or T" and "false or F." Propositional logic is based on Boolean algebra based on the lattice $L=\{\mathbf{0},\mathbf{1}\}$ or, equivalently $\{T,F\}$. One very general way of introducing Boolean logic is by looking at its underlying algebra as special kind of lattice. A lattice is an ordered set of elements containing with each pair of elements their least upper bound called *supremum* and the greatest lower bound called *infimum*. Two binary operations are defined on this lattice, these are: the join (V) and meet (Λ) operations. A Boolean algebra (Boolean lattice) denoted by $\langle \mathbf{L}, \vee, \wedge, \neg, \mathbf{0}, \mathbf{1} \rangle$ is a lattice L which is ordered and distributive and has a complement operation. The Boolean algebra for this so-called classical logic is then: $\mathcal{L}_{\mathcal{B}} = \langle \{\mathbf{0},\mathbf{1}\}, \vee, \wedge, \neg \rangle$. This Boolean algebra of classical logic is isomorphic to a Boolean algebra of crisp sets. Boolean logic and its underlying two-valued algebra play a predominant role in the social science analyses, if not always openly recognized as such. For example, the neatness of the dichotomy of truth values is inscribed at the heart of Boolean analysis of causality, thereby equipping the study of causality with a number of powerful features (Braumoeller, 2003). First, it clearly explicates the logic of combining different independent variables. Second, it makes dealing with complex causality a straightforward matter. One can clearly delineate how independent variables jointly or independently determine the outcomes. One can also use Boolean analysis to explicate the possibility of substitutability among causal variables. Third, Boolean analysis clearly and consistently explicates necessary and sufficient conditions as well as various combinations thereof. Fourth, the process of Boolean minimization helps to systematically isolate relevant from irrelevant independent variables (Ragin, 1987:85-163). Fifth, it also includes counterfactual analysis through the negation operator which strengthens the validity of the causal analysis.

In formal logic, atomic or elementary propositional variables are combined into formula using the logical connectives: Implication $\rightarrow$, conjunction $\wedge$, disjunction $\vee$, negation $\neg$, and equivalence $\leftrightarrow$. Using two propositional variables or formula **P** and **Q** we can obtain new formula such as $\mathbf{P \rightarrow Q}$, $\mathbf{P \wedge Q}$, $\mathbf{P \vee Q}$, $\mathbf{\neg P}$, $\mathbf{P \leftrightarrow Q}$. The truth values of the compound formula are obtained from a combination of the truth values of its elements using the following basic Truth Table:

		negation	Implication	Equivalence	Conjunction	Disjunction
P	Q	$\neg P$	$P \rightarrow Q$	$P \leftrightarrow Q$	$P \wedge Q$	$P \vee Q$
T	T	F	T	T	T	T
T	F	F	F	F	F	T
F	T	T	T	F	F	T
F	F	T	T	T	F	F

Boolean logic forces us to think only in terms of true and false statements. Vagueness, which is an inherent feature of human linguistic concepts, is simply thrown away. However, as the wealth of statistical and qualitative studies show, vagueness is inescapable in all our theoretical and empirical analyses of social and political phenomena. Statements that such and such variable causes more or less such and such outcome are pervasive in all social sciences. What is important here is that this vagueness is not only a type of epistemic uncertainty that might be resolved with better tools of analysis. Rather, vagueness of this type goes deep down into the logic underpinning our analyses. This led Lotfi Zadeh to propose in the 1960s fuzzy logic as a solution to this inescapable problem of vagueness. Zadeh was preceded for many years by other scholars who suggested multi-valued (MV) logic as a generalization to two-valued Boolean logic. What we really need in social sciences is to explore the implications of these MV and fuzzy logics for the logical bases of our social science inquiries.

The key idea in many-valued logic is that the truth value of a proposition can assume other values than just "True" or "False." When this truth value is numerically represented it is usually assumed that it belongs to the interval of real numbers [0,1], where 0 corresponds to completely false and 1 to completely true. This has many implications some of which nullify a number of laws drawn from Boolean logic such as the law of the excluded middle. In Boolean logic a statement can either be true or false, hence a proposition like the law of excluded middle **(P∨¬P),** i.e., for every proposition **P,** either **P** or **not-P** holds, is consistent and exists. An equivalent law of Boolean logic is *reductio ad absurdum* or proof by contradiction: For every proposition **P**, if **not-P** does not hold, then **P** holds. Contra Boolean logic, fuzzy logic and MV logics in general do allow for many values between true and false. A basic premise of MV logics is that it is incorrect to think that to know the world it is sufficient to know the "truths," we also need to know the "falsities," as well as the degrees of both. There are many situations in our daily lives where the Boolean logic is simply not practically relevant. Consider a simple example from ordinary language usage when one answers a yes-or-no question by simply stating: Yes and No. Mutually inconsistent scientific theories are another illustration of this phenomenon. In sum, relaxing the principle of bivalence of truth and falsity and allowing for the possibility of one or more truth values between true and false leads to many possible logics with many new truth values.[2]

2 Linguistic Fuzzy Variables

Analysis based on linguistic fuzzy logic is mathematically rigorous and coherent, built on a coherent mathematical framework of definitions, theorems, lemmas and the like.[3] In this chapter I present a brief introduction of the linguistic fuzzy logic framework. In this pursuit, the first step is to choose the basic ingredients that are

---

[2] So-called four-value Belnap's logic is one such example with the semantics underlying the logic is based on the set of truths: {T, F, Both, None}.

[3] Herrera and Herrera-Viedman (2000).

used in the symbolic manipulation. This means that we need to choose a context-dependent linguistic terms-set to describe vague or imprecise information. For example, a linguistic terms set for the linguistic variable *Feasible* can be defined as: *F(Feasible)={fully feasible, very feasible, approximately feasible, possibly feasible, approximately feasible or possibly feasible, approximately feasible and possibly feasible, not feasible}*.

Normally, in any one situation, one is faced with a number of linguistic variables, not just one. It is easier to opt for the same linguistic terms set to describe variation of the various linguistic concepts in a given problem, although this is not a requirement. The elements in the set will determine the level of distinction among different parts of the available information, called the granularity of vagueness. Each value of the linguistic variables is characterized by a *syntactic label* and a *semantic value* or *meaning*. The label is a word or a sentence belonging to the chosen linguistic terms set. The meaning of the label is a fuzzy subset in a universe of discourse (a finite set of words and/or phrases). The choice of the linguistic terms set with its semantic is thus the first step of any linguistic approach to solving a problem. A linguistic variable is mathematically defined as follows:

<u>*Linguistic Variable:*</u>

A linguistic variable is a quintuplet (L, S(L), U, G, M). L is the name of the linguistic variable. S(L) denotes the term set of L, i.e., the set of names that provide the linguistic values of L, with each value being a fuzzy variable denoted generally by X and ranging across a universe of discourse U. G is a syntactic rule (which usually takes the form of a grammar) for generating the names of the values of L. M is a semantic rule for associating meanings M(L) to L, with M(L) being a subset of U.

This definition of a linguistic variable can practically be implemented by means of an ordered structure of linguistic terms. The semantic of these linguistic terms is hence derived from their own ordered structure. The cardinality of the terms set must be small enough so as not to impose useless precision on the users. Yet, it ought to be rich enough to allow enough discrimination of the information being analyzed. Typical values of cardinality used in the literature on linguistic models are odd ones, usually varying from 7 to 13. The mid linguistic term usually represents an assessment of a statement such as "approximately 0.5," with the rest of the terms being placed symmetrically around it (Figure 1).

{NNN=Null, VLL=Very_Low, LLL=Low, MMM=Medium, HHH=High, VHH= Very_High, FFF=Full}

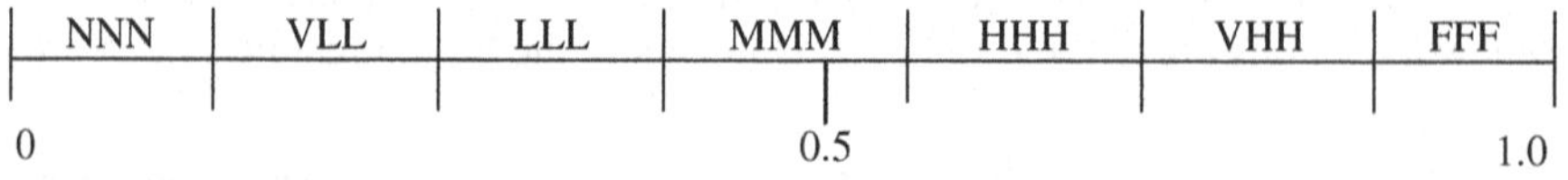

Fig. 1 A symmetrically distributed ordered set of seven linguistic terms

Using the example shown in Figure 1 a set of seven terms S used to describe various levels of cooperation could be given as:

S-cooperation = {s_0 = *Null*, s_1 =*Very Low*, s_2 = *Low*, s_3 = *Moderate*, s_4 = *High*, s_5 = *Very High*, s_6 = *Full*}-*cooperation*

This set is ordered in the sense that $s_a \prec s_b$ if, and only if, a < b.

As an illustration let us consider one of the example that Charles Ragin (2000) examines in his book – IMF protest. The study seeks to understand the causes of IMF protest in various countries, with some having more protests than others. Ragin categorizes countries on how severe an IMF protest they are confronted with. This dependent variable is explained using a model with several independent variables, namely, IMF pressure, Urbanization, Economic hardship, Dependence on investment, Political liberalization, and Government activism. Not all countries face IMF protest and those that do face it at different levels. Hence, argues Ragin, IMF protest is a rather fuzzy category or set, with some countries more included in it than others. Using linguistic terms, we can say that some countries are *fully* in the category of countries facing IMF protest, others are mostly but no fully in the category, others are mostly but not fully out the category, and others are fully out of the category of countries confronted with IMF protest. Ragin considers seven levels of membership into the category of "IMF protest": *fully in, mostly but not fully in, more or less in, neither in nor out, more or less out, mostly but not fully out, and fully out.* I denote these respectively as: FLIN, MNFI, MLIN, NINO, MLOU, MNFO, FLOU. Using these linguistically expressed partial memberships, called linguistic values, Ragin's data for the dependent variable IMF protest (IMFP) reads as shown in Table 1. For reason of notational consistency with the remaining part of the paper I have converted Ragin's scheme of linguistic values of the dependent variable into a new one – called in this paper *nuance value* – as shown in the fourth and fifth columns of Table 1. For example, Argentina has a Full membership (PRT = PPP) in the category of IMF protest while Costa Rica has a Low membership (PRT = LLL) in the category of IMF protest.

Ragin also considers the independent variables – *IMF pressure (IMP), Urbanization (URB), Economic hardship (ECH), Dependence on investment (DEI), Political liberalization (POL), and Government activism (GVA)* – as fuzzy categories to which the different countries belong to a certain level using the same partial membership levels or level of belongingness.

From Ragin's Table 10.1 we read, for example, that Argentina is fully included in the set of IMF protest (PRT = FLIN = PPP), mostly but not fully included in the set of countries facing IMF pressure (IMP = MNFI = VHH), fully included in the set of countries that are urbanized (URB = FLIN = PPP), fully included in the countries that face economic hardship (ECH = FLIN = PPP), more or less out of the set of the countries that depend on investment (DEI = MLOU = LLL), fully in the set of countries that politically liberalized (POL = FLIN = PPP), and fully in the set of countries with government activism (GVA = FLIN = PPP). (Note that

Table 1 Linguistic Values of the Dependent Variable PRT (= IMF Protest) for various countries.

Dependent Variable: IMF Protest: PRT					
	Ragin's Value		Nuance Value		
Numerical Value	Symbol	Meaning	Symbol	Meaning	Country
1.00	FLIN	Fully In	PPP	Full	Argentina, Peru
0.83	MNFI	Mostly but Not Fully In	VHH	Very-High	Brazil
0.67	MLIN	More or Less In	HHH	High	Bolivia, Chile
0.50	NINO	Neither In Nor Out	MMM	Moderate	Dominican Republic, Ecuador, Haiti, Jamaica, Morocco, Poland, Sudan, Yugoslavia, Zaire
0.33	MLOU	More or Less Out	LLL	Low	Costa Rica, Egypt, India, Nigeria, Philippines, Romania, Sierra Leone, Tunisia, Turkey, Venezuela, Zambia
0.17	MNFO	Mostly but Not Fully Out	VLL	Very-Low	El Salvador, Ghana, Guatemala, Hungary, Ivory Coast, Jordan, Liberia, Mexico, Niger, Panama
0.00	FLOU	Fully Out	NNN	Null	

the first linguistic value is that of Ragin's original notation and the second one is mine).[4]

In Boolean logic, stating that a variable acquires a certain linguistic value implies that the truth value of this assignment is "True." An assignment of truth value is always one of two choices – True or False. The situation is radically different in linguistic fuzzy logic. In addition to possessing a linguistic value (such as *big*, *small*, *extremely low*, or *approximately high*), each value assignment to a linguistic variable has a degree of truth. For example, a country A with a low membership value to the category of countries facing IMF protest, i.e., PRT = LLL, could have a *very-high, moderate, low,* or *very low* degree of truth. In other words, our attribution of a low membership for country A to category PRT is true to a *very-high, moderate, low,* or *very low* degree. The notion of truth degree thus plays the role of a linguistic level of confidence in the estimated nuances.

Manipulating linguistic variables using linguistic fuzzy logic thus introduces two "degrees of freedom," one dealing with the linguistic value of the variable termed in this book as *nuance value* (or level) and one dealing with the truth value of this assigned nuance value termed as *truth (or confidence) degree* (or value). Instead of just speaking of a certain level of membership in a category, we speak of *a level of nuance*, that is, the variable is *more or less nuanced.* We say, for example, a country is a very-high authoritarian. We then pose the question: How true is each of (or, equivalently, how much confidence do we have in) these value assignments of nuance? Is *lower nuance* truer than *higher nuance*? In other words, each of these degrees of nuance is *more or less true*. This is a sort of a fuzzification of the membership functions of fuzzy-set theory. *The membership degrees (into different fuzzy categories) are also fuzzy sets*. The linguistic fuzzy

[4] I do this to remain consistent in using the same system of notation throughout the whole book.

Table 2 Example of LFLA version of PRT (= IMF Protest) for various countries.

Dependent Variable: IMF Protest: PRT				
Nuance Value		Truth Value		
Symbol	Meaning	Symbol	Meaning	Country
PPP	Full	PPP	Full	Argentina, Peru
VHH	Very-High	LLL	Low	Brazil
HHH	High	VLL	Very-Low	Bolivia, Chile
MMM	Moderate	VHH	Very-High	Dominican Republic, Ecuador, Haiti, Jamaica, Morocco, Poland, Sudan, Yugoslavia, Zaire
LLL	Low	LLL	Low	Costa Rica, Egypt, India, Nigeria, Philippines, Romania, Sierra Leone, Tunisia, Turkey, Venezuela, Zambia
VLL	Very-Low	VHH	Very-High	El Salvador, Ghana, Guatemala, Hungary, Ivory Coast, Jordan, Liberia, Mexico, Niger, Panama
NNN	Null	NNN	Null	

logic approach presented in this book thus involves a type-2 fuzzification of the variables.[5]

At this point of the discussion, it is already clear that the LFLA proposed in this book differs from Ragin's fuzzy-sets approach in two respects. First, I do not use numerical values of membership functions. I instead consider the degrees of belongingness to a category as a linguistic variable expressed as nuance value. Second, I introduce a second type of fuzziness at a deeper level than what Ragin considers – the membership degrees are also fuzzy sets, or more exactly, fuzzy linguistic terms. Using LFLA, Ragin's data on the category of IMF protest (PRT) would for example take the form shown in Table 2. I am thus assuming various truth values for the different linguistic nuance values of Variable PRT. For example, Argentina and Peru have a full degree of membership in the category of IMF protest (PRT = PPP) which is known with a full degree of truth (PPP), whereas Brazil has very high degree of membership (PRT = VHH) which is known with a low degree of truth (LLL). The point is that this second level introduces another type of fuzziness (vagueness) in the data that Ragin does not consider.

In order to rigorously account for the notions of nuance and truth values and various systematic manipulations thereof we need a systematic procedure – an algebraic structure – capable of allowing that, as explained next.

3 Algebra of Linguistic Nuance and Truth Values

The algebraic structure of a system of logic is a formal framework for carrying out a rigorous analysis using linguistic fuzzy logic. To see this point more clearly let us consider the following set of possible linguistic values: *T*={*true*, *false*, *very true*, *very false*, *approximately true*, *possibly true*, *approximately true or possibly*

[5] This approach corresponds to type-2 fuzzy-sets theory as known in artificial intelligence literature where the membership functions are also numerical fuzzy sets, whereas Ragin's approach uses what is known as type-1 fuzzy-sets theory (see: Mendel, 2001).

true, approximately true and possibly true}. Suppose that we have two linguistic variables A and B with the value of A being *approximately true* and that of B being *possibly true*. What would be the linguistic value of A AND B? What would be the value of A OR B? Would these respectively be *approximately true and possibly true, approximately true or possibly true*? The axiomatic system of an algebraic structure would allow us to make such evaluations.

The axiomatic system of linguistic values is based on the idea of using hedges (or modifiers) to change the meaning and values of basic elements such as: True and False. For example, using the hedge *very* we would obtain two new truth values, *very true* and *very false*. The obtained algebra is called hedge algebra, that is, it is an algebraic structure that defines and determines how the hedges are combined with the two basic truth values to form a complete system of evaluation. This algebra is based on natural semantic properties of linguistic hedges, the set of which is equipped with an ordering relation. To make the system of nuance and truth values a practical and consistent way of evaluating the linguistic values of causal inferences we need to equip the set of linguistic values with a joint operation, a meet operation, a negation operation, and an implication operation (Ho and Khang, 1999; Ho and Nam, 2002).

Linguistic hedges are very useful because they possess a number of semantic properties which make their algebraic manipulation intuitively meaningful. First, each linguistic primary term such as true and false possesses a semantic tendency which can intuitively be expressed as a semantically ordering relation. In addition, one primary term would possess a meaning which is stronger (higher) than another one. Thus, we can say that *true* $\succ$ *false*. Moreover, these terms have a tendency to be semantically positive or negative. "*True*" is positive in the sense that *very true* $\succ$ *true* and "*false*" is negative in the sense that *very false* $\prec$ *false*. However, it is also possible to have a reversal of the meaning of the primary term in the sense that we would for example have *approximately true* $\prec$ *true*. One point to note here is that these orderings are intuitively sound in natural language. Second, hedges are semantic modifiers with different degrees of modification. This makes it possible to compare the effects of two hedges on the same primary term. Hence, we would have *little* $\prec$ *approximately* since *little true* $\prec$ *approximately true*. Third, hedges change the meaning of the primary term, but nevertheless do preserve an original essential meaning of the primary term – this is called semantic heredity. For example, from *little true* $\prec$ *approximately true* we have *possibly little true* $\prec$ *possibly approximately true*. Let us now move on to the formulation of an abstract hedge-algebra.

Let us denote the set of truth-hedges by $T_M = \{\tau_\alpha, \alpha \in [1, M]\}$ where $\tau_\alpha - true$ is a symbolic truth value (with a similar formalism for linguistic nuance degrees). We would for example have the following set of hedges for truth values: T_9={*not at all, little, enough, fairly, moderately, quite, almost, nearly, completely*}. The truth values will be: {*not at all true, little true, enough true* , *fairly true, moderately true, quite true, almost true, nearly true, completely true*}. In order to be able to manipulate symbolically (not numerically) these truth values we need to provide the set of truth-hedges with an algebraic structure. We call this an hedge algebra and denote it by **HA=**$\{T_M, \vee, \wedge, \prec, \rightarrow, \neg\}$ where $\vee$ and $\wedge$ are respectively the disjunction and conjunction, $\rightarrow$ is the implication, $\neg$ is the negation, and $\prec$ is an ordering of the truth-hedges (nuance-hedges). We take **HA** to be a De Morgan lattice, satisfying the following definition.

Definition: de Morgan Lattice:

Let **L** = {**M**, $\vee$, $\wedge$, <, $\neg$} be a lattice. *L is called a De Morgan lattice if:*

1. {**M**, $\vee$, $\wedge$, <} *is a distributive lattice*
2. *The operation* $\neg$: **M** $\rightarrow$ **M** *is such that:*
   a. *For all* **u** $\in$ **M** *we have:* $\neg\neg$ **u=u**
   b. *For all* **u, v** $\in$**M** *we have:* $\neg(u\wedge v) = \neg u \vee \neg v$ **and** $\neg(u \vee v) = \neg u \wedge \neg v$.

De Morgan lattices are distributive lattices with a complement operation which satisfies the law of double complement, the De Morgan rules, and the law of contraposition for partial ordering <. If a De Morgan lattice is bounded with a smallest element **0** and a largest element **1** in the partial ordering then it is called a De Morgan algebra, with the property that $\neg$**0=1** and $\neg$**1=0.** For example, if the lowest element of **HA** is "*null*" the corresponding highest element would then be "*full.*"

In numerical many-valued logic, one conventional way to define the conjunction (AND) and disjunction (OR) operations $\wedge$ and $\vee$ is by **a**$\wedge$**b=min(a,b)** and **a**$\vee$**b=max(a,b)**. The disjunction and conjunction operations $\vee$ and $\wedge$ in linguistic many-valued logic can be defined using two equivalent operators, namely, LWD (linguistic weighted disjunction) and LWC (linguistic weighted conjunction).

***Linguistic Conjunction*: AND: a$\wedge$b = LWC (a,b)**
***Linguistic Disjunction:* OR: a$\vee$b = LWD (a,b)**

These two operators are special cases of a more general form called LWA (linguistic weighted averaging operator) discussed below. The three operators LWD, LWC, and LWA as well as the linguistic implication LI are defined in terms of a more basic operator LOWA (linguistic ordered weighted average) as well as linguistic versions of MAX and MIN operators. I next introduce LWD, LWC, LWA, and LI operators (Herrera and Herrera-Viedman, 1997).

4 Linguistic Aggregation Operators

A key point in formalizing the symbolic manipulation of linguistic variables is that using these variables calls for two sorts of aggregating operations of antecedents which we need to consider in carrying out LFLA analysis through disjunction, conjunction, implication, negation, and various combinations thereof. First, aggregation of truth values – each variable has its own linguistic truth value. Hence, we need to aggregate these linguistic truth values to obtain the collective truth value for the aggregate. This is done using the LOWA aggregation (as explained shortly). Second, aggregation of weighted degrees of nuance – we need to combine the nuance degrees of the variables with different weights according to the type of logical combination being considered (disjunction, conjunction, etc...). This can be done using the operators LWC (linguistic weighted conjunction), LWD (linguistic weighted disjunction), LWA (linguistic weighted average), and LI (linguistic implication). As an illustration, let us consider for example the following hypothesis about war termination (Chan, 2003).

H1: Wars with a 'small and slow' start tend to last longer than wars with a 'big and fast' start.

Let us recast H1 using the conceptual tools being developed in this section in the following way:

1. For a given dyad of states – α, we know that {War starts with a level of fighting LOF (e.g., *more or less small*) and a rate of fighting ROF (e.g., *more or less slow*)} and we know this to a degree of truth τ_α (e.g., *moderately true*).
2. We know the duration of this war DOW (e.g., *moderately big start*) and the degree of truth of this knowledge is $\tau_{\alpha-DOW}$ (e.g., *almost true*).
3. What is the degree of truth τ_β of the statement that {If Wars start with a small level of fighting [LOF=small] at a slow rate of fighting [ROF=slow] then they would tend to have a longer war duration [DOW=longer] than Wars that start with [LOF=big] and [ROF=fast]}?

Put differently: we have in statement 3 a logical implication: **A → B**? That is: We know the weighted nuance degrees of both A and B and we know the truth values of both A and B, can we infer the nuance and truth values of the implication **A → B**? Phrased in this way we are testing for fuzzy causal necessity. The nuance value characterizes how fuzzy this causal implication is and the truth value tells us how confident we should be in this estimate. In order to test for causal sufficiency we need to consider the statement: **¬A → ¬B**? In solving this problem we need two procedures: a first one on how to aggregate the linguistic information on the nuance levels of DOW for both A and B and a second one on how to aggregate the linguistic information on the truth values of the DOW variables for both A and B. Aggregating the linguistic information on A and B consists as discussed earlier of two steps – (1) aggregation of individual truth values and (2) aggregation of weighted linguistic degrees of nuance of A and B. In comparison, Chan's analysis based on Boolean logic is as follows: First, DOW is either small or big, and hence either relevant or not relevant – the degree of nuance of this variable is either 1 or 0. Second, the truth value of DOW is "true." In Boolean logic, knowledge about the degree of nuance corresponds to a truth value of "true." No knowledge corresponds to a truth value of "false." In the approach suggested in this book,

DOW can have different degrees of nuance and each of these degrees of nuance has a linguistic truth value. That is, if we know that for example DOW is low-to-moderate nuance in the analysis we still need to inquire about the truth values of this knowledge about the nuance of DOW.

For the sake of defining the four operations – LWC, LWD, LWA, LI – let us denote by v_m the nuance degree of a linguistic variable L_m and τ_m its truth value $-L_m = (v_m, \tau_m)$. It is clear that we need to posit that the set of linguistic truth values and the set of nuance degrees of the variables have the same cardinality – each linguistic variable has one degree of nuance and one truth value. Moreover, without any loss of generality we assume that both v and τ are linguistically expressed using the same set $S = \{s_0, s_1, \cdots, s_T\}$ of linguistic, ordered labels, with cardinality $T+1$. We have the following definitions (Herrera and Herrera-Viedman, 1997).

<u>*Linguistic Neg, MAX, and MIN*</u>

$$\text{Negation Operator: } Neg(s_i) = s_{T-i} \tag{1}$$

$$\textit{Maximization Operator: } MAX\left(s_i, s_j\right) = \left\{s_i \text{ if } i \geq j;\ s_j \text{ otherwise}\right\} \tag{2}$$

$$\textit{Minimization Operator: } MIN\left(s_i, s_j\right) = \left\{s_i \text{ if } i \leq j;\ s_j \text{ otherwise}\right\} \tag{3}$$

As an illustration consider the set: O_7={Null, Very Low, Low, Moderate, High, Very High, Full}. This set is ordered since we can intuitively see that Null ≤ Very Low ≤ Low ≤ Moderate ≤ High ≤ Very High ≤ Full. The negation operator gives: Neg(Null) = Full and Neg(Low) = High, etc. The maximization operator gives: MAX(Null, Moderate) = Moderate and MAX(Very High, Moderate)=Very High, etc. The minimization operator gives: MIN(Low, High) = Low and MIN(Full, High) = High, etc.

<u>*LWC Aggregation*</u>

The LWC_ω aggregation of m linguistic variables $\left\{(v_1, \tau_1), \cdots, (v_m, \tau_m)\right\}$ *is defined by:*

$$\left(v_{LWC}, \tau_{LWC}(\omega)\right) = LWC_\omega\left\{(v_1, \tau_1), \ldots, (v_m, \tau_m)\right\} \tag{4}$$

The aggregated nuance value is

$$v_{LWC} = MIN_{i=1,\ldots,m}\left\{MAX\left(Neg\left(v_i\right), \tau_i\right)\right\} \tag{5}$$

The aggregated truth degree is

$$\tau_{LWC}(\omega) = LOWA_\omega\left(\tau_1, \ldots, \tau_m\right) \tag{6}$$

The LOWA$_\omega$ aggregation – linguistic ordered weighted aggregation – depends on the orness parameter ω and is defined (and illustrated through an example) down below. As an example of using LWC let us consider:

$$\left(v_{LWC}, \tau_{LWC}\right) = LWC\left[\left(LLL, VHH\right), \left(HHH, LLL\right)\right] \tag{7}$$

We want to aggregate two linguistic variables A=(LLL,VHH) and B=(HHH,LLL) with, respectively, a degree of nuance LLL for A and HHH for B, a degree of truth VHH for A and LLL for B. The degree of nuance of the aggregate is obtained as:

$$v_{LWC} = MIN\left\{MAX\left(Neg\left(LLL\right), VHH\right); MAX\left(Neg\left(HHH\right), LLL\right)\right\} \tag{8}$$

Using the hedge algebra {NNN, VLL, LLL, MMM, HHH, VHH, PPP}, we have: $Neg(LLL) = HHH$ and $Neg(HHH) = LLL$. Thus:

$$v_{LWC} = MIN\left\{MAX\left(HHH, VHH\right); MAX\left(LLL, LLL\right)\right\} = MIN\left\{VHH; LLL\right\} = LLL \tag{9}$$

The degree of truth is obtained as an aggregation of the respective degrees of feasibility for A and B; that is: $\tau_{LWC} = LOWA\left(VHH, LLL\right)$.Using LOWA as given down below we obtain:

$$\tau_{LWC} = LOWA\left(VHH, LLL\right) = MMM\ . \tag{10}$$

Hence:

$$\left(v_{LWC}, \tau_{LWC}\right) = (LLL, MMM) \tag{11}$$

The aggregated variable has a Low degree of nuance with a Moderate degree of truth.

<u>*LWD Aggregation*</u>

The LWD$_\omega$ aggregation of m linguistic variables $\left\{\left(v_1, \tau_1\right), \cdots, \left(v_m, \tau_m\right)\right\}$ *is defined by:*

$$\left(v_{LWD}, \tau_{LWD}\left(\omega\right)\right) = LWD_\omega\left[\left(v_1, \tau_1\right), \cdots, \left(v_m, \tau_m\right)\right] \tag{12}$$

The aggregated nuance value is

$$v_{LWD} = MAX_{i=1,\ldots,m}\left\{MIN\left(v_i, \tau_i\right)\right\} \tag{13}$$

The degree of truth of the aggregated variable is

$$\tau_{LWD}\left(\omega\right) = LOWA_\omega\left(\tau_1, \ldots, \tau_m\right) \tag{14}$$

LWA Aggregation

The aggregation of the set of m weighted individual variables $\{(\nu_1,\tau_1),\cdots,(\nu_m,\tau_m)\}$ *according to the* LWA_ω *operator is defined by*

$$(\nu_{LWA}(\omega),\tau_{LWA}(\omega)) = LWA_\omega\left[(\nu_1,\tau_1),\cdots,(\nu_m,\tau_m)\right] \quad (15)$$

The collective nuance degree of the aggregated variable is

$$\nu_{LWA}(\omega) = LOWA_\omega(\nu_1,\cdots,\nu_m) \quad (16)$$

The collective truth value is

$$\tau_{LWA}(\omega) = LOWA_\omega\left[MIN(\nu_1,\tau_1),\cdots,MIN(\nu_m,\tau_m)\right] \quad (17)$$

LI Operation[6]

The linguistic fuzzy logic implication LI_ω *is defined through the* LWD_ω *operator by:*

$$(\nu_{LI},\tau_{LI}(\omega)) = LI_\omega\left[(\nu_i,\tau_i),(\nu_j,\tau_j)\right] = LWD_\omega\left[Neg\{(\nu_i,\tau_i)\};(\nu_j,\tau_j)\right] \quad (18)$$

Where:

$$\nu_{LI} = MAX\left\{MIN\left[Neg(\nu_i,\tau_i)\right];MIN\left[(\nu_j,\tau_j)\right]\right\} \quad (19)$$

$$\tau_{LI}(\omega) = LOWA_\omega\left(Neg(\tau_i),\tau_j\right) \quad (20)$$

5 Aggregation of Linguistic Information

Several kinds of aggregation operators of linguistic information have been suggested in the literature.[7] They can be grouped in four broad categories: (1) aggregation operators of linguistic non-weighted information, (2) aggregation operators of linguistic weighted information, (3) aggregation operators of multi-granularity linguistic information, and (4) aggregation operators of numeric and linguistic information. These operators differ according to the kind of information that is to be aggregated. Type (1) is used when the different pieces of linguistic information to be aggregated are all equally weighted in the process of decision analysis. Type (2) is most useful when some pieces of linguistic information are

[6] This is a generalization of the usual definition of S-implication as: **a → b ≡ ¬aVb**.

[7] Delgado et al., 1998; Herrera, Herrera-Viedma, and Verdegay, 1996; Herrera, Herrera-Viedma, and Martinez, 2000; Herrera and Herrera-Viedman, 2000.

more important than others. Both types (1) and (2) are situations where the granularity – semantic differentiation – of the various pieces of information is the same. Type (3) applies to situations where different pieces of linguistic information have different semantics such as, for example, one having 7 terms in the linguistic terms set whereas the other having 13 terms. Although this is not a fundamental difficulty, it does make the manipulation of linguistic labels somewhat more demanding. Finally, type (4) is important in situations when linguistic and quantitative pieces of information are combined together in the decision making process. Again this does not pose a fundamental obstacle but does increase the number of steps that one has to make in analyzing the decision process. For reasons explicated down below I adopt the second method for aggregating linguistic information.

This type of aggregation of linguistic information is done by using a LOWA or Linguistic Ordered Weighted Averaging operator. Two essential steps in this aggregation are: first, an ordering to the linguistic labels and, second, weighted convex combination of the labels two by two until we aggregate all labels together. As an illustration of this procedure, suppose that we would like to aggregate the following four labels, {*L, ML, H, VH*}. This is a subset of a larger set

$$S=\{s_0,\cdots,s_8\}=\{VL,L,ML,FFML,FFMH,MH,H,VH\}$$

We have:

VL=Very_Low,

L=Low,

ML=Medium_Low,

FFML=From_Fair_to_More_or_less_Low,

F=Fair,

FFMH= From_Fair_to_More_or_less_High,

MH=More_or_less-High,

H=High, VH=Very_High.

First, we need to decide on the weights that we attach to each label in the combination. That is, we need to have four weights, i.e., a weighing vector W=[.,.,.,.]. Below I explicate the procedure through which we can obtain these weights depending on the characteristics of the problem at hand. Let's assume for the sake of this illustration the following weighting vector W=[0.3,0.2,0.4,0.1]. Because these are the weights for an ordered set, the first weight corresponds to the "highest" label in the linguistic set {*L,ML,H,VH*}, in this case VH, and so on. The convex combination (after ordering the set of labels) is thus symbolically written as:

$$\Phi(L,ML,H,VH)=[0.3,0.2,0.4,0.1](L,ML,H,VH)=C^4\{(0.3,VH);(0.2,H);(0.4,ML);(0.1,L)\} \tag{21}$$

At this stage, we have yet to start the process of aggregation. The aggregation is done in two phases:

1. We symbolically decompose $\Phi(L,ML,H,VH)$ into a convex combination of simpler aggregations of labels (that is, with a smaller number of labels to aggregate) until we reach the smallest convex combination, that is, a combination of two labels.[8]
2. We go backward and re-compose until we reach a final unique label. This is done as follows.

We begin by decomposing C^4 as a convex combination of the highest label (in this case VH) which has a weight of 0.3 and the aggregation of the remaining labels with the remaining weight 0.7, that is,[9]

$$\begin{aligned}&C^4\{(0.3,VH);(0.2,H);(0.4,ML);(0.1,L)\}=\\&0.3\otimes VH\oplus(1-0.3)\otimes C^3\{(0.29,H);(0.57,ML);(0.14,L)\}\end{aligned} \tag{22}$$

The new weights 0.29, 0.57, and 0.14 for the labels H, ML, and L inside C^3 (combination of three labels) are obtained by dividing the corresponding original weights in C^4 (combination of four labels) by the sum of the three original weights of H, ML, and L, that is,

0.29=0.2/(0.2+0.4+0.1) as the new weight for H;
0.57=0.4/(0.2+0.4+0.1) as the new weight for ML;
0.14=0.1/(0.2+0.4+0.1) as the new weight for L.

The sum for the new weights is 0.29+0.57+0.14=1, as it should be for a convex combination. We next apply the same procedure to C^3, thus getting

$$C^3\{(0.29,H);(0.57,ML);(0.14,L)\}=0.29\otimes H\oplus(1-0.29)\otimes C^2\{(0.8,ML);(0.2,L)\} \tag{23}$$

The new weights

0.80=0.57/(0.57+0.14) as the new weight for ML;
0.20=0.14/(.57+0.14) as the new weight for L.

And we repeat the operation with C^2 which leads to:

$$C^2\{(0.8,ML);(0.2,L)\}=0.8\otimes ML\oplus(1-0.8)\otimes L \tag{24}$$

---

[8] A fundamental aspect of this operator is the re-ordering step, in particular an aggregate *A* is not associated with a particular weight *W* but rather a weight is associated with a particular ordered position of the aggregate.

[9] $\otimes$ and $\oplus$ are used to signify different symbolic operations.

In sum, we have the following:[10]

$$C^4\{(0.3,VH);(0.2,H);(0.4,ML);(0.1,L)\}=$$
$$0.3\otimes VH \oplus (1-0.3)\otimes\{0.29\otimes H \oplus (1-0.29)\otimes[0.8\otimes ML \oplus (1-0.8)\otimes L]\} \quad (25)$$

At this stage we use the following rule for combining two labels:

$$\begin{cases} w_1 \otimes s_j \oplus (1-w_1)\otimes s_i = s_k & \text{with always} \quad j \geq i \\ k = \min\{T, i + round(w_1 \cdot (j-i))\} & T \text{ is the cardinality of } S \\ round(x) = \begin{cases} [x] & if\ |x-[x]| \leq 0.5 \\ 1+[x] & otherwise \end{cases} \end{cases} \quad (26)$$

$[x]$ is the integer part of x. Note that in the case of $m=2$ we have $b_1 = s_j$ and $b_2 = s_i$. If $w_j = 1$ and $w_i = 0$ with $i \neq j, \forall i$, then the convex combination is defined as $C^2\{w_i, b_i, i = 1,2\} = b_j$.

We thus obtain for the above example $k = \min\{8, 1 + round(0.8\cdot(2-1))\} = \min\{9,2\} = 2$ since ML= s_2 and L= s_1, that is,[11]

$$C^2\{(0.8,ML);(0.2,L)\} = 0.8\otimes ML \oplus (1-0.8)\otimes L = ML \quad (27)$$

In order to arrive at the final result we need to go backward by successively retracing the previous steps, that is, by recombining the labels. Hence, for $m=3$

[10] This procedure is possible because the LOWA operator can be shown to satisfy the three properties of "increasing monotonicity," "associativity," and "commutativity." This implies that we can carry out the decomposition in a variety of different ways which would all lead to the unique final result.

[11] Symbolic methods might suffer a "loss" of information caused by the use of the *round* operator, which comes from the need to express the results in the initial expression domain which is discrete and finite. This problem of loss of information can be corrected by using the so-called 2-tuple symbols (a, S) where S is a label such as H, VH , etc., and a $\in$[-0.5,0.5[. This procedure becomes important when [x]≠x (i.e., x is not an integer number). In this case a=x-[x] helps us decide which nearest label to choose. I do not use these finessing techniques in this paper to avoid an additional burden on the reader. See Herrera and Martinez (2000) for details.

$$C^3\{(0.29,H);(0.57,ML);(0.14,L)\}=0.29\otimes H\oplus(1-0.29)\otimes ML=FFML(=s_3) \tag{28}$$

Because H=s_7 and ML= s_2, we thus find *min{9,2+round(0.29(7-2))}=min{9,3}=3*. For *m=4*

$$C^4\{(0.3,VH);(0.2,H);(0.4,ML);(0.1,L)\}=0.3\otimes VH\oplus(1-0.3)\otimes FFML(=s_3) \tag{29}$$

since VH= s_8 and FFML=s_3 and thus *min{9,3+round(0.3(8-3))}=min{9,5}=5*. The final result is:

$$\Phi(L,ML,H,VH)=MH\ . \tag{30}$$

In words: Combining the labels Low, Medium_Low, High, and Very_High with weights (0.3,0.2,0.4,0.1) produces a label with value More_or_Less_High. More generally, we have the following definition of a Linguistic Ordered Weighted Operator – LOWA.[12]

Definition of LOWA Operator:

Let S = {s_1,…, s_m} be a set of m labels to be aggregated. The LOWA operator, Φ, is defined as:

$$\Phi(s_1,\ldots,s_m)=W\otimes B^T=C^m\{w_k,b_k;k=1,\ldots,m\}=w_1\otimes b_1\oplus(1-w_1)\otimes C^{m-1}\{\beta_h^{m-1},b_h;h=2,\ldots,m\}$$

where $\beta_h^{m-1}=w_h/\sum_2^m w_k$ with h=2,…,m.

$W=\{w_1,\ldots,w_m\}$ *is a weighing vector satisfying: (i)* $w_i\in[0,1]$ *and (ii)* $\sum_i^m w_m=1$.

$B=\{b_1,\ldots,b_m\}$ *is a vector associated to S such that* $B=\sigma(S)=\{s_{\sigma(1)},\ldots,s_{\sigma(m)}\}$ *is an ordering of S in which* $s_{\sigma(i)}\le s_{\sigma(j)}$ *for* $j\le i$*, with σ being a permutation over the set of labels S.*

C^m *is called the convex combination operator of m labels. In the case of m=2 it is defined by:* $C^2\{w_i,b_i,i=1,2\}=w_1\otimes s_j\oplus(1-w_1)\otimes s_i=s_k$ *with* $s_j,s_i\in S$ *such that* $j\ge i$.

k is obtained as $k=\min\{T,i+round(w_1\cdot(j-i))\}$.

[12] The LOWA operator possesses a number of properties which makes the process of aggregation of linguistic information a rational one. These properties are: unrestricted domain, unanimity or idempotence, positive association of social and individual values, independence of irrelevant alternatives, citizen sovereignty, and neutrality (Herrera et al., 1996).

Note that if $w_j = 1$ and $w_k = 0$ for all $k \neq j$, then $C^m\{w_k, b_k; k = 1,...,m\} = b_j$. The weights $W = \{w_1,...,w_m\}$ make LOWA an "orand" operator located somewhere between the "AND" and "OR" logical operations. In other words, LOWA has simultaneously a finite degree of "orness" and a finite degree of "andness," defined by:

$$andness(W) = 1 - orness(W) \tag{31}$$

$$orness(W) = \frac{1}{n-1}\sum_{j=1}^{n}(n-j)w_j \tag{32}$$

If, for example, $W=[0.4,0.3,0.2,0.1]^T$ then *orness*(W)=(1/3)(3(0.4)+2(0.3)+1(0.2)) =0.666. The meanings of *orness* and *andness* operators are clearly seen in the following cases:

$$\begin{aligned}
&Maximum\ Operator: \quad W^{MAX} = [1,0,0,0,...,0]^T \quad orness(W^{MAX}) = 1 \quad andness(W^{MAX}) = 0 \\
&Minimum\ Operator: \quad W^{MIN} = [0,0,0,0,...,1]^T \quad orness(W^{MIN}) = 0 \quad andness(W^{MIN}) = 1 \\
&Average\ Operator: \quad W^{AVE} = [\tfrac{1}{n},\tfrac{1}{n},...,\tfrac{1}{n}]^T \quad orness(W^{AVE}) = 0.5 \quad andness(W^{AVE}) = 0.5
\end{aligned} \tag{33}$$

Expressed in terms of fuzzy sets the maximum operator (producing the highest linguistic label in the ordered set) corresponds to an intersection of the fuzzy sets of all linguistic labels, and hence picks up the smallest common set, which in this case will be the highest label. This produces a degree of *orness* of 1 and a degree of *andness* of 0. Conversely, expressed in terms of fuzzy sets the minimum operator (producing the lowest linguistic label in the ordered set) corresponds to a union of the fuzzy sets of all linguistic labels, and hence picks up the largest common set, which in this case will be the lowest label – the lowest label is necessarily implied in all subsequent labels.[13] This produces a degree of *orness* of 0 and a degree of *andness* of 1. The average operator does not differentiate between all labels and hence produces equal degrees of *orness* and *andness*, i.e., 0.5. Hence, LOWA operator with many of the weights near the top will be an *orlike* operator with $orness(W) \geq 0.5$, while those operators with most of the weights at the bottom will be *andlike* operators with $orness(W) \leq 0.5$.

To offer a more intuitive understanding of the LOWA operator and the role of the weighting factors let me draw a parallel with the body of literature on decision theory using the poliheuristic method as developed by Alex Mintz and others (Mintz, 1993; 1995; Mintz and DeRouen, 1994; Mintz and Geva 1997).[14] Briefly, the poliheuristic approach posits five main characteristics of decision making. (1)

[13] This is possible because the labels are a complete ordered set.

[14] See the special issue of *The Journal of Conflict Resolution* (2004) 48 (1).

Non-holistic search: The process of decision making derives not from "evaluation and comparison of all alternatives across different dimensions" but rather from the use of "heuristic decision rules that do not require detailed and complicated comparisons of relevant alternatives, and adopts or rejects undesirable alternatives on the basis of one or a few criteria" (Mintz and Geva, 1997: 85). (2) Dimension or criterion-based processing: In situations where decision-makers are faced with cognitive and/or environmental constraints, they tend to use a dimension or criterion-based process instead of an alternative-based approach for processing information (Redd, 2004:339). (3) Non-compensatory decision rules: When decision makers are faced with multiple goals under constraints they resort to a non-compensatory strategy which "avoids alternatives that have radically different values in key goals because, to compensate for a low value, one needs a high one on other dimensions" (Goertz, 2004: 16). Hence, the approach emphasizes "how leaders evaluate different alternatives in light of their multiple and often conflicting aims. This emphasis poses a sharp contrast with standard expected utility models that most often make assumptions about the form of the one-dimensional utility function" (Goertz, 2004:15). (4) Satisficing behavior: Ambiguity, uncertainty, and cognitive and other constraints end the decision making process not through maximization of options but rather when an acceptable alternative is found to satisfy a number of key criteria (Mintz and Geva, 1997: 86-87). (5) Order-sensitive search: Most theories of decision making (based on some variants of expected utility theory) posit the invariance assumption, that is, regardless of how a decision task is framed with respect to the ordering and sequencing of dimensions and alternatives, the outcome should remain the same. The poliheuristic approach argues that the invariance assumption is violated for, as put by Mintz and Geva (1997: 87), "the poliheuristic decisionmaking model implies that the choice of a particular alternative may depend on the order in which particular dimensions (diplomatic, economic, military, political) are invoked." In a nutshell, the LOWA method exactly does what the poliheuristic approach suggests, that is, LOWA offers one way to formally model the poliheuristic theory approach.

From a poliheuristic theory perspective the LFLA is non-holistic, aggregates information based on a limited set of criteria, allows a non-compensatory aggregation of the information on the various alternatives, is based on a satisficing strategy, and is order sensitive. Aspects (1) and (2) of the poliheuristic theory method are built in the LOWA operator as defined up above, whereas aspects (3), (4), and (5) are taken care of through the aggregation weights W. Different choices of W would allow different degrees of non-compensatory, satisficing and sensitivity to ordering in the decision making process. Hence, it can be said that the weights W are in some sense a modeling and measure of these three aspects of the poliheuristic theory method. In this respect, the degree of *andness* (as defined earlier) would provide a measure of the degree of non-compensatoriness in decision making (while *orness* would measure the degree of compensatoriness). Moreover, as explained in the next chapter, the LFLA also offers one way to model the two-stage procedure of decision-making postulated by the poliheuristic theory method. Yet, the LFLA is more general than the poliheuristic method for it can as well formally model a decision making process based on different assumptions concerning the five aspects of poliheuristic theory (1–5) discussed earlier. Determining the weights W differently formally models different methods of aggregating information such as for example allowing for different degrees of compensatoriness.

The LOWA aggregation of labels depends crucially on the weights vector. Yager (1988) introduced entropy as a measure that indicates the degree to which *W* takes into account the various individual pieces of information (expressed by the linguistic labels). Entropy measures the uncertainty on the occurrence of a distribution of weights. Probability is traditionally used to measure the uncertainty that we have on a single event, whereas, in comparison, entropy measures the uncertainty on a collection of events. A highly concentrated distribution of weight has smaller entropy than a less concentrated distribution; the former distribution is more informative and it is easier to predict its outcomes. Entropy provides a measure of information uncertainty in the sense that it describes the difficulty of predicting an outcome of a random variable.[15] Entropy for a weighting vector W is defined as:

$$E(W) = -\sum_{i=1}^{n} w_i \ln w_i$$

Entropy is an interesting quantity because it satisfies the so-called *Maximum Entropy Principle* (MAXENT). The intuition behind the MAXENT is that the W that maximizes the entropy is a reasonable estimate of the true distribution when we lack any other type of information. The estimate will be altered if we receive new information. The result is that out of all possible estimates of the probability distributions that are consistent with the "data", the MAXENT method picks the one which is most uninformative; that is, closest to a uniform distribution (Golan, Judge, and Karp; 1996; Paris and Vencovskfa, 1996). MAXENT thus stipulates that in the process of making inferences based on incomplete information, we should select the probability distribution function with maximum entropy value given the observed data (Jaynes, 1982). Hence, the distribution function obtained using MAXENT provides the most conservative estimation of the unknown underlying distribution function. This choice singles out the most significant and least biased distribution and the one which best represents the true distribution (Jaynes, 1957; 1989).[16]

As an example imagine that you were to choose from a choice-set the elements of which are not necessarily fair (i.e., a non-uniform distribution of choices) defined by throwing two non-fair dices. There are eleven possible outcomes, each

---

[15] Information theorists use entropy as a measure of the average amount of information required to describe the distribution of some random variable of interest (Shannon, 1948; 1949; Cover and Thomas 1991).

[16] For illustrative works using the notion of entropy in political science see, for example, Theil, 1969; Gill, 1997; in economics, Ullah, 2002; in sociology, Alexander, 1996; in game theory, O'Connell and Stearns, 2003; Neyman, 1999. For a comparison with Bayesian approaches, see, for example, Shen and Perloff, 2001.

outcome having unknown probabilities of occurrence. Even though you do not know the distribution of outcomes, it is better for you to bet on the number 7 as opposed to 2 or 12 since there are 6 combinations that produce 7 while there is only 1 for the other combinations. The bet on 7 is the maximum entropy bet because the combination 7 produces the maximum expected value (given the lack of prior information about the probabilities). Thus, the probability distribution which maximizes the entropy is numerically identical with the frequency distribution which can be realized in the greatest number of ways. The MAXENT principle means that making parametric inferences based on incomplete information should be done by selecting the probability distribution function with the maximum entropy value given the observed data.

MAXENT principle could also be interpreted as a criterion for selecting a probability model. As long as we are willing to view entropy as a suitable measure of uncertainty, MAXENT principle allows us to select the distribution which contains the largest amount of uncertainty compatible with certain constraints. This is somewhat similar to the maximum likelihood principle with the important difference that in the case of MAXENT one does not have to a priori assume an analytical form for the distribution. MAXENT principle is thus an important generalization of Laplace's principle of insufficient reason which stipulates that the uniform distribution in situations where there is complete lack of knowledge about the events. MAXENT deals with partial information and hence stipulates that we should choose the probabilities that maximize the uncertainty about the missing information.

In sum, using MAXENT in the aggregation process of linguistic fuzzy information is suitable since it falls well within the spirit of preserving as much vagueness and uncertainty as possible. When supplemented with the measure of *orness* we have a procedure which preserves the vagueness of the information used to reach a decision as well as the multi-dimensionality character of the decision making process. Thus, I combine the two measures – *orness* and entropy – to determine the weights at every level of *orness* using MAXENT by solving the following optimization problem:

$$\text{Maximize} \qquad E(W) = -\sum_{i=1}^{n} w_i \ln w_i$$

$$\text{Subject to:} \qquad orness(W) = \omega = \frac{1}{n-1}\sum_{i=1}^{n}(n-i)w_i$$

$$w_i \in [0,1] \ (1 \le i \le n)$$

$$\sum_{i=1}^{n} w_i = 1$$

where ω is the specified desired degree of *orness*.

An analytic solution to this optimization problem can be found using the Lagrange multiplier method.[17] Let α and β be two Lagrange multipliers. The method of Lagrange multipliers consists in optimizing the following expression:

$$L = -\sum_{i=1}^{n} w_i \ln w_i + \alpha\left(\sum_{i=1}^{n} w_i - 1\right) + \beta\left(\frac{1}{n-1}\sum_{i=1}^{n}(n-i)w_i - \omega\right) \tag{34}$$

The partial derivatives of L are given by:

$$\begin{aligned}
\frac{\partial L}{\partial w_i} &= -\ln w_i - 1 + \alpha + \beta\frac{n-i}{n-1} = 0, \quad 1 \le i \le n \\
\frac{\partial L}{\partial \alpha} &= \sum_{i=1}^{n} w_i - 1 = 0, \\
\frac{\partial L}{\partial \beta} &= \frac{1}{n-1}\sum_{i=1}^{n}(n-i)w_i - \omega = 0.
\end{aligned} \tag{35}$$

Simultaneously solving these three equations, we obtain:

$$w_i = \frac{h^{n-i}}{\sum_{j=1}^{n} h^{n-j}} \tag{36}$$

h is related to the multiplier β by $\beta = (n-1)\ln h$. h is the largest positive solution of the following equation:

$$\sum_{i=1}^{n}\left(\frac{n-i}{n-1} - \omega\right)h^{n-i} = 0 \tag{37}$$

For every degree of *orness* ω we obtain a value of h which gives a weights vector W. For example, for n=3 and a desired degree of *orness* $\omega = 0.7$, we obtain the following equation:

$$(1.-0.7)h^2 + (0.5-0.7)h + (0-0.7) = 0.3h^2 - 0.2h - 0.7 = 0 \tag{38}$$

The positive solution of this equation is h=1.89, leading to $w_1 = 0.55; w_2 = 0.30; w_3 = 0.15$ with an optimized entropy of $E(W)$=0.97. Table 3 shows the solutions for a system with n=4 weights for various degrees of *orness*.

[17] Filev and Yager, 1995; Fuller and Majlender. 2001.

Table 3 Weights Vector using MAXENT Principle

orness	***andness***	h	w_1	w_2	w_3	w_4	***E(W)***
1.00	0.00		1.00	0.00	0.00	0.00	0.00
0.96	0.04	9.29	0.892	0.096	0.010	0.001	0.38
0.79	0.21	2.27	0.581	0.256	0.112	0.049	1.06
0.66	0.44	1.49	0.413	0.276	0.185	0.124	1.29
0.50	0.50	1.00	0.250	0.250	0.250	0.250	1.38
0.34	0.66	0.67	0.124	0.185	0.276	0.413	1.29
0.21	0.79	0.44	0.049	0.112	0.256	0.581	1.06
0.04	0.96	0.10	0.001	0.010	0.096	0.892	0.38
0.00	1.00	0.00	0.000	0.000	0.000	1.000	0.00

Chapter 3
Linguistic Fuzzy-Logic Decision-Making Process

The chapter presents a linguistic fuzzy-set approach to decision making that models how actors who are faced with situations fraught with vagueness make their decisions. The chapter considers situations where the actors engage in a process of evaluating the merits of different choice alternatives under multiple criteria – a process of Multiple Criteria Decision Making. A key question of this chapter is: how do actors aggregate the vague information that they have on these criteria to reach a final decision on the situation facing them? Drawing on a vast area of research in engineering and other fields of knowledge such as business management and medical research, I propose a linguistic fuzzy-logic approach to analyze the process of decision making in social and politics under multiple criteria.[1]

The chapter is organized into three sections. A first section introduces a hypothetical example to be used as illustrative anchor for explicating the model. A second section explicates a linguistic fuzzy theory approach to the process of decision-making under conditions where the decision makers are required to simultaneously satisfy multiple criteria, which might be mutually reinforcing or conflicting. A third section compares the linguistic fuzzy set approach to social choice theory. This section hence helps make the case that the linguistic fuzzy set method can be applied to individual decision making as well as collective choice situations. The hypothetical situation is one where state leaders have to decide on the formation of a security alignment by choosing among a set of alternative arrangements based on a set of multiple criteria. This example is meant to be illustrative only and hence I do not claim any empirical "truth" to the underlying substantive assumptions or conclusions.

1 International Security as Illustration

Let us assume a hypothetical situation where state leaders are seeking to form a security alignment. In order to do so they need to weigh a number of alternative forms of security alignments taking into account a number of factors (or criteria). Let's assume that in order to choose a form of security alignment state leaders

[1] For some illustrative literature, see: Herrera and Herrera-Viedma, 2000, and references therein; Bordogna and Passi, 1993; Delgado et. al., 1998; Gurocak and Whittlesey, 1998; Herrera and Rodriguez, 2002; Herrera et. al, 2001; Herrera-Viedma, 2001.

B. Arfi: Linguistic Fuzzy Logic Methods in Social Sciences, STUDFUZZ 253, pp. 43–62.
springerlink.com

have to consider simultaneously four factors: DDR, VUL, RPL, and RMT (see Table 1 for the meanings of these symbols).[2] State leaders engage in a process of Multiple Criteria Decision Making (MCDM) in evaluating the merits of different alternatives among a set of possible security alignments.

I call these variables criteria following Zadeh (1970: B 148) who demonstrated that in a fuzzy set approach "concepts of goal and constraint ... are defined as fuzzy sets in the space of alternatives and thus ... can be treated identically in the formulation of a decision." The crisp separation between goals and constraints disappears, and the fuzzy goals and constraints are aggregated to a single function that is maximized.[3] This insight has been much elaborated on in the 30 years following this and is now widely accepted in the milieu of fuzzy set theorists.[4] Along these lines, criteria and conditions are treated in the same way in my chapter. To make the story short, whether we can call these criteria, conditions, or simply factors – what is important, and this is a value-added of the fuzzy-sets approach, is how they are aggregated to produce a result. The reason for this is that vagueness can never be eliminated from these and most human concepts and this is precisely what the fuzzy set approach is about; that is, how to model this vagueness and not force an artificial crispness on natural language. A fuzzy logic approach – instead of Boolean logic approach – does not require, in fact discourages, seeking a sharp delineation between criteria and conditions. For example, the argument is that when a practitioner is seeking to reach a decision on whether to conclude an alliance pact he/she does not make a list of criteria and a second list of conditions and then treats them differently in his/her mind. Rather, the practitioner tries his/her best to account for he/she thinks are the most important factors influencing the issue at hand.

I thus posit that the state leaders need to choose among a set of four alternative options of security alignment $\{WBB,SBB,MUU,SCC\}$ (see Table 1 for explanation of these symbols). They do this by combining the available linguistic information on the four factors $\{DDR,VUL,RPL,RMT\}$. A state leader who is interested in forming a security alignment with other states comparatively evaluates the four options by taking into account all four factors.

The variables $\{DDR,VUL,RPL,RMT\}$ are considered as linguistic variables. We hence need a semantic set for expressing their variations. I choose the following set of terms $\{NNN,LLL,LMM,MMM,MHH,HHH,VHH\}$ where *NNN* = *None*, *LLL*= *Low*, *LMM*= *Low_ to_Medium*, *MMM*= *Medium*, *MHH*= *Medium_to_High*, *HHH*= *High*, and *VHH*= *Very_High*. Thus, in an arrangement of

[2] This hypothetical example has strong affinity with David Lake's (1999) theory on entangling relations. One key difference though is that Lake does not include mutual trust as an additional independent variable that determines the type of security alliance formed.

[3] It is worthwhile recalling here Herbert Simon's (1996) statement, "In the decision-making situations of real life, a course of action, to be acceptable, must satisfy a whole set of requirements, or constraints. Sometimes one of these requirements, or constraints, is singled out and referred to as the goal of the action. But the choice of one constraint from many is to a large extent arbitrary. For many purposes it is more meaningful to refer to the whole set of requirements as the (complex) goal of the action. This conclusion applies both to individual and organizational decision-making."

[4] See, for example, Herrera and Herrera-Viedma, 1998.

Table 1 Symbols used in the security alignment example

Security Alignment		Factors (or Criteria)	
Symbol	**Meaning**	**Symbol**	**Meaning**
WBB	Weak Bilateralism	**DDR**	**D**egree of **D**iffuse **R**eciprocity among allied states
SBB	Strong Bilateralism	**VUL**	**VUL**nerability among allied states to abandonment, entrapment, and/or exploitation
MUU	Multilateralism	**RPL**	**R**eliance of **P**ower **L**everage among allied states
SCC	Security Community	**RMT**	**R**eliance of **M**utual **T**rust among allied states

Weak Bilateralism (*WBB*), I posit that the two states expect that the shared degree of diffuse reciprocity is low (*DDR=LLL*). A low *DDR* creates a fear of high mutual vulnerability (*VUL=HHH*). If there is a situation of power asymmetry the strongest would then seek to use its power leverage to tilt the mutual security relation in its favor. That is, the powerful would readily rely very much on its power leverage (*RPL*) in defining and maintaining the arrangement. But then this would mean that the two states would not rely much on mutual trust (*RMT*) in defining and maintaining the bilateral arrangement. In this situation we thus have *RPL=VHH* and *RMT=LLL*. Following the same type of reasoning, I posit Table 2 which summarizes the linguistic values of the four factors {*DDR*,*VUL*,*RPL*,*RMT*} for each of the four possible choices of security arrangement.[5] Let me emphasize that a key aspect of Table 2 is not the absolute fuzzy values of the individual variables {*DDR*, *VUL*,*RPL*,*RMT*} for each of the security options {*WBB*,*SBB*,*MUU*,*SCC*}. Rather, what is hypothesized and yet to be demonstrated is that the aggregation of the values of these individual variables does indeed produce the posited security option. Hence, one "hidden" weakness of this analysis is that Table 2 is stated in comparative static terms. Yet, the choice of security alignment is the result of the combined effects of all four variables, {*DDR*,*VUL*,*RPL*,*RMT*}.

A key question thus is: how to aggregate the effects of the individual variables to check the validity of the results of comparative static analysis? One conventional way to do it is by estimating the value of each variable, let's say for weak bilateralism, and then estimate the overall costs and benefits that a state might incur were it to opt for such a security arrangement. The deciding state would also have to carry out the same cost/benefit analysis for other forms of security alignment. The final choice on the form of security arrangement would then be the one that minimizes the overall costs and maximizes the overall benefits. The final choice for any state would be obtained by optimizing the estimate of the aggregated values of all four variables. Yet, this approach is still underdetermining for it does not explicitly address the above "hidden" weakness; it does not demonstrate explicitly the postulated simultaneous multiple causality.

I propose to use a linguistic fuzzy-logic framework to model how state leaders make the choices that they do. In doing this I am positing that the decision makers do not seek first to estimate the absolute costs and benefits of the independent

[5] This table is simply posited for the sake of illustration – I am not suggesting that there is a substantive theory underpinning these assignments.

variables and then only at the end compare the relative merits of the aggregated costs and aggregated benefits of all security arrangements. Rather, state leaders consider the relative merits of the security arrangement right from the beginning. For example, a state leader would examine how WBB fares on the dimension of strategic vulnerability when compared with multilateralism, and so on and so forth for all four variables $\{DDR,VUL,RPL,RMT\}$. Only at the end do state leaders consider the aggregated relative merits of different security arrangements. I believe that this is very realistic for it does not require the analyst (or the policy maker) to estimate the absolute values of the independent variables. This is a crucial point for the task of estimating the absolute values of variables is always fraught with methodological pitfalls and empirical hurdles that all students of IR and practitioners alike are well aware of. Not only the full range of alternative security arrangements must be specified and evaluated according to their relative merits. I also think it appropriate and realistic that we should deal with the independent variables in the same way. A LFLA framework allows us to do this using fuzzy comparison. In other words, I take it that the results displayed in Table 2 are not yet fully demonstrated – they have yet to **simultaneously** take into account the multiple criteria. This step is more often assumed implicitly by most scholars who engage in this kind of analysis. Demonstrating multiple causality is often not done, or in most cases rather vaguely asserted. The approach does indeed provide one way to establish multiple causality (or lack thereof) in a formal and rigorous mathematical way using linguistic fuzzy analysis.

Table 2 Security Arrangements

	DDR	***VUL***	***RPL***	***RMT***
WBB?	*LLL*	*HHH*	*VHH*	*LLL*
SBB?	*LMM*	*VHH*	*HHH*	*LMM*
MUU?	*HHH*	*VHH*	*LLL*	*HHH*
SCC?	*VHH*	*VHH*	*NNN*	*VHH*

More formally, the analysis consists of asking a number of IF-Then fuzzy questions to reach a conclusion about the combination of the different dimensions of causality. In the case of the running example we would have for WBB the following fuzzy rule:

IF {"DDR≈LLL" © "VUL≈HHH" © "RPL≈VHH" © "RMT≈LLL"} **THEN** WBB

I use the © symbol instead of AND or OR because in the LFLA multiple causality is neither strictly necessary (which is usually represented with a logical AND and symbolically by a product of the independent variables) nor strictly sufficient (which is usually represented with a logical OR and symbolically by an addition of the independent variables).[6] LFLA allows a formalization of fuzzy multiple causality as "more or less necessary like" and "more or less sufficient

[6] See for example: Goertz (2003a; 2003b:68-70) and Chapters 6 and 7 of this book.

like." The outcome WBB is causally produced through the fuzzy rule IF-THEN to a certain extent, which is expressed by a linguistic variable. In conventional decision making causality is usually treated using Boolean two-valued logic phrased formally as **WBB=DDR*VUL*RPL*RMT** if, for example, all independent variables are considered to be necessary conditions. Alternatively, we could have **WBB=DDR*VUL* RPL+DDR*VUL*RMT** if, for example, DDR and VUL and RPL are jointly necessary conditions and DDR and VUL and RMT are jointly necessary conditions (and a number of other possible combinations). If all four variables are sufficient conditions we can write in Boolean logic of causality **IF DDR, VUL, RPL, RMT then WBB**. For probabilistic causality this would take the form **IF DDR, VUL, RPL, RMT then probably WBB**. It is important to note that with fuzzy IF-THEN rules we are dealing with vague causality not with probabilistic causality, that is, fuzzy causality means that WBB **IS produced to a partial degree**[7] whereas probabilistic causality would mean that WBB **IS produced OR NOT produced with a given likelihood**. Hence, a key difference between LFLA decision making and conventional decision making (such as used in expected utility) is that they are built on different notions of causality rooted in different logics.[8]

The first task now is to establish how each of the four options $\{WBB, SBB, MUU, SCC\}$ compares to one another according to each of the four

Table 3 Semantic Set for Preference Relations

k	*Label*	*Linguistic Meaning*
12	*DPP*	x_i is preferred to x_j in **D**efinite degree
11	*VHP*	x_i is preferred to x_j in **V**ery **H**igh degree
10	*HPP*	x_i is preferred to x_j in **H**igh degree
9	*MPP*	x_i is preferred to x_j in **M**oderate degree
8	*LPP*	x_i is preferred to x_j in **L**ow degree
7	*VLP*	x_i is preferred to x_j in **V**ery **L**ow degree
6	*AAS*	x_i is about the same **AS** x_j
5	*VLD*	x_j is preferred to x_i in **V**ery **L**ow degree
4	*LDD*	x_j is preferred to x_i in **L**ow degree
3	*MDD*	x_j is preferred to x_i in **M**oderate degree
2	*HDD*	x_j is preferred to x_i in **H**igh degree
1	*VHD*	x_j is preferred to x_i in **V**ery **H**igh degree
0	*DDD*	x_j is preferred to x_i in **D**efinite degree

7 To put it in an awkward way: WBB IS "a bit caused" AND "a bit not caused;" that is, causality is partial.

8 At a fundamental or logic level the difference is that LFLA deals with partial degrees of truth whereas probabilistic causality still adheres to a Boolean notion of truth with a dichotomous choice between TRUE and FALSE (at least for most works published in social sciences).

variables $\{DDR, VUL, RPL, RMT\}$. These estimates are expressed linguistically and are relative, not absolute. To move further, we need to introduce a second semantic set that can be used to express how these security options perform compared to one another when taking into account the various criteria. I opt for the following set which is big enough to allow a rich comparison of the four security options.[9]

Each label corresponds to an index k which is used for purposes of structurally ordering the labels in the set. The whole set corresponds to $\{p_k \mid k = 0,\ldots,12\}$ with each element of the set representing a preference relation. Table 3 is set up to provide dyadic comparisons of security options x_i and x_j. Using these semantics we obtain 14 different 4x4 matrices of performance relations, with each matrix corresponding to a linguistic value of one of the four criteria $\{DDR, VUL, RPL, RMT\}$. For example, for $VUL=VHH$, I obtain the following matrix

$$\boldsymbol{VUL=VHH}, \quad P^1 = \begin{bmatrix} - & LPP & LPP & LPP \\ LDD & - & AAS & AAS \\ LDD & AAS & - & AAS \\ LDD & AAS & AAS & - \end{bmatrix}$$

Elements 1,2,3,4 of a row correspond to the order in the set $\{WBB, SBB, MUU, SCC\}$ (and the same for column elements). This means that on the dimension of strategic vulnerability, we have the following preference relations between the four forms of security arrangement (*ceteris paribus*):

When mutual vulnerability (**VUL**) is very high (**VHH**):

$P^1_{12} = LPP$ ⇔ *WBB is preferred to SBB in Low degree*

$P^1_{13} = LPP$ ⇔ *WBB is preferred to MUU in Low degree*

$P^1_{14} = LPP$ ⇔ *WBB is preferred to SCC in Low degree*

$P^1_{21} = LDD$ ⇔ *WBB is preferred to SBB in Low degree*

$P^1_{23} = AAS$ ⇔ *SBB is about the same as MUU*

$P^1_{24} = AAS$ ⇔ *SBB is about the same as SCC*

$P^1_{31} = LDD$ ⇔ *WBB is preferred to MUU in Low degree*

$P^1_{32} = AAS$ ⇔ *MUU is about the same as SBB*

$P^1_{34} = AAS$ ⇔ *MUU is about the same as SCC*

$P^1_{41} = LDD$ ⇔ *WBB is preferred to SCC in Low degree*

$P^1_{42} = AAS$ ⇔ *SCC is about the same as SBB*

$P^1_{43} = AAS$ ⇔ *SCC is about the same as MUU*

[9] Note that the preference relations for *k=0,…,5* are the mirror images of those for *k=7,…,12*.

The remaining 13 matrices are listed in Appendix A. These matrices are derived from the assumptions made in Table 2. I first take as starting points the linguistic values ascribed to the four variables {*DDR,VUL,RPL,RMT*} and, second, derive the comparison relations of the four options {*WBB,SBB,MUU,SCC*} for each of these linguistic values. For example, in the case of the matrix for *VUL=VHH* displayed earlier, the *WBB* option is the more likely one to be chosen, that is, weak bilateralism is the most preferred security option, *ceteris paribus*. Yet, state leaders know that vulnerability is only one of four criteria to be satisfied and therefore they must take into account the other criteria as well. Under such a constraint, state leaders would, for example, compare *WBB* with *SCC* and decide that under a condition of *VUL=VHH* they would rank *WBB* as $P_{14}^{1} = LPP$ compared to *SCC* (that is *WBB* is preferred to *SCC* to a Low degree), *ceteris paribus*. Along the same lines, state leaders would however rank *SBB* as about the same ($P_{24}^{1} = AAS$) as *SCC*, the reason being that under a condition of *VUL=VHH*, *SBB* is closer to *SCC* than *WBB* on the dimension of *VUL* (as shown in Table 3), *ceteris paribus*. Again, these evaluations are expected to be rough estimates only. This is the essence of using a Linguistic fuzzy approach to the problem of decision making – *preserving a lack of precision is important*. The remaining preference relations are obtained through a similar process of reasoning.

As a second illustration of this procedure let us consider the case of SNF (short range nuclear forces) decision that occurred within NATO in 1989 which Sanjian (1992) examined using a fuzzy-set approach. NATO faced four different options: X1 – modernize SNF, X2 – negotiate missile reduction with the Warsaw Treaty Organization, X3 – modernize SNF and negotiate with WTO, X4 – neither modernize nor negotiate by postponing the decision on SNF. The preferences ranking for the members of NATO (as deduced by Sanjian) are listed in Table 4.

Table 4 Preference orderings on the SNF issue (Table I from Sanjian (1992:279)).

State	*Ordering*	*State*	*Ordering*
Belgium	X2 > X4 > X3 > X1	Luxembourg	X2 > X4 > X3 > X1
Canada	X1 > X3 > X4 > X2	Netherlands	X3 > X1 > X4 > X2
Denmark	X2 > X4 > X3 > X1	Norway	X2 > X4 > X3 > X1
France	X3 > X4 > X1 > X2	Portugal	X3 > X1 > X4 > X2
Germany	X2 > X4 > X3 > X1	Spain	X2 > X4 > X3 > X1
Greece	X2 > X4 > X3 > X1	Turkey	X3 > X1 > X4 > X2
Iceland	X2 > X4 > X3 > X1	United Kingdom	X1 > X4 > X3 > X2
Italy	X2 > X4 > X3 > X1	United States	X1 > X3 > X4 > X2

Using a method based on membership functions, Sanjian predicts that Germany's preference ordering is what NATO will most likely adopt, which indeed was the case as the empirical evidence shows.

The NATO collective choice can alternatively be analyzed using the linguistic fuzzy approach. To this end, we need to recast the information in Table 4 in terms of a linguistic terms-set and produce matrices of preference relations that compare for each state the various options as dyads. Although I could use the linguistic set introduced earlier, there is no need for such a high level of granularity and I instead opt to use the following reduced set of linguistic terms $\{HPP, MPP, LPP, AAS, LDD, MDD, HDD\}$ (as defined in Table 5). This gives a total of 16 preference matrices.

Table 5 Semantic Set for Preference Relations for expressing NATO members' preferences

k	*Label*	*Linguistic Meaning*
6	*HPP*	x_i is preferred to x_j in **H**igh degree
5	*MPP*	x_i is preferred to x_j in **M**oderate degree
4	*LPP*	x_i is preferred to x_j in **L**ow degree
3	*AAS*	x_i is about the same **AS** x_j
2	*LDD*	x_j is preferred to x_i in **L**ow degree
1	*MDD*	x_j is preferred to x_i in **M**oderate degree
0	*HDD*	x_j is preferred to x_i in **H**igh degree

Let us consider, for example, the German case for which Sanjian (1992) derives the following ordering of alternative options from best to worst option: X2 > X4 > X3 > X1. This is translated using linguistic preference relations as X2 is preferred to X1 in high degree (HPP); X2 is preferred to X3 in moderate degree (MPP); X2 is preferred to X4 in low degree (LPP). Likewise, X4 is preferred to X1 in moderate degree (MPP); X4 is preferred to X3 in low degree (LPP). And X3 is preferred to X1 in low degree (LPP). This leads to the following matrix of comparative preferences for Germany:

$$\{X2, X4, X3, X1\} \Rightarrow P^{Germany}\{X1, X2, X3, X4\} = \begin{bmatrix} - & HDD & LDD & MDD \\ HPP & - & MPP & LPP \\ LDD & MDD & - & LDD \\ MPP & LDD & LPP & - \end{bmatrix}$$

The row elements of this matrix corresponds to the vector {X1,X2,X3,X4}, whereas the column elements correspond to the transposed vector $\{X1,X2,X3,X4\}^{\text{transpose}}$. The matrix preferences for the other 15 states are obtained in a similar fashion.

$$P^{Canada} = \begin{bmatrix} - & HPP & LPP & MPP \\ HDD & - & MDD & LDD \\ LDD & MPP & - & LPP \\ MDD & LPP & LDD & - \end{bmatrix}; \quad P^{France} = \begin{bmatrix} - & LPP & MDD & LDD \\ LDD & - & HDD & MDD \\ MPP & HPP & - & LPP \\ LPP & MPP & LDD & - \end{bmatrix}$$

$$P^{Netherlands} = \begin{bmatrix} - & MPP & LDD & LPP \\ MDD & - & HDD & LDD \\ LPP & HPP & - & MPP \\ LDD & LPP & MDD & - \end{bmatrix}; \quad P^{UK} = \begin{bmatrix} - & HPP & MPP & LPP \\ HDD & - & LDD & MDD \\ MDD & LPP & - & LDD \\ LDD & MPP & LPP & - \end{bmatrix}$$

The remaining matrices are given by:

$$P^{Belgium} = P^{Debnmark} = P^{Greece} = P^{Germany} = P^{Iceland} = P^{Italy} = P^{Luxembourg} = P^{Norway} = P^{Spain}$$

$$P^{Turkey} = P^{Netherlands} = P^{Portugal}$$

$$P^{US} = P^{Canada}$$

In order to reach a decision on what type of security arrangement is best for a state or to reach a collective decision in the NATO example, the information on dyadic preference relations has to be aggregated to be useful. The procedure for doing so is addressed next.

2 Choosing the Best Alternative

The task is to choose the best alternative among a set of n linguistic options $X = \{x_1, ..., x_n\}$ according to m linguistic criteria $C = \{c_1, ..., c_m\}$. The starting point is to evaluate the various options in X in comparative terms, that is, how an alternative performs in satisfying each of the multiple criteria compared to the other options. This process produces a set of performance relations matrices, P_i, made up of performance relations, with a matrix per each criterion i. For m criteria we would have m (n x n) matrices $\{P_1, ..., P_m\}$ of performance relations. The goal then consists in finding the best performance relations matrix. The latter is found by aggregating the linguistic information stored in the individual performance relation matrices through the LOWA operator introduced in Chapter 2 according to the following two steps (Herrera, Herrera-Viedma, and Verdegay, 1996):

1. The different linguistic degrees of preference relations are evaluated for each alternative x_i according to the set of m criteria considered individually. This produces the individual degrees of preference, ID_i^k, for each alternative x_i according to each criterion k.
2. The different linguistic degrees are evaluated for each alternative according to the set of criteria as a whole by aggregating the individual degrees of preference, ID_i^k. This produces the social degrees of preference, SD_i for each alternative x_i.

The actual process of choosing the best alternative is done through a process called of linguistic dominance among a set of decision options in three steps (see Figure 1).

1. For each matrix P^k of linguistic preference relations according to criterion k, we use the LOWA operator with a specified *weights vector* W_1 (determined by using MAXENT principle for a degree of *orness* ω_1), denoted as Φ_{W_1}, to obtain the individual linguistic dominance degree of each alternative x_i, ID_i^k, as $ID_i^k = \Phi_{W_1}\left(p_{ij}^k, j = 1,\ldots,n; j \neq i\right)$ with $k=1,\ldots,m$; and $i=1,\ldots,n$.
2. For each alternative x_i, we then evaluate the social linguistic dominance degree, SD_i, using a second LOWA operator with a specified *weights vector* W_2 (determined by using MAXENT principle for a degree of *orness* ω_2), denoted as Φ_{W_2}, as $SD_i = \Phi_{W_2}\left(ID_i^k, k = 1,\ldots,m\right)$ with $i=1,\ldots,n$.
3. We obtain the set of alternatives that have maximum linguistic dominance degrees $X_{\max}^d$ as $X_{\max}^d = \left\{x_i \in X \middle| SD_i = \max_j\left(SD_j\right)\right\}$.

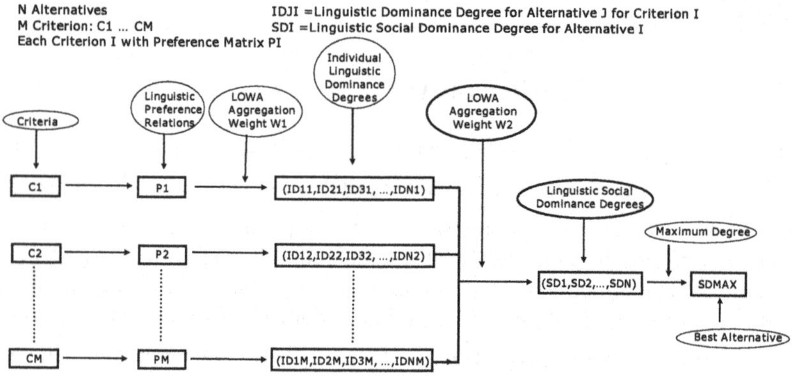

Fig. 1 Two-Level LOWA Aggregation Process

This method of choosing the best alternative is now illustrated in the running example on security alignment. I assume a degree of *orness* of 0.3 at the first step of aggregation (that is, individual linguistic dominance degree of each alternative) and a degree of *orness* of 0.5 at the second step of aggregation (that is, social linguistic dominance degree of each alternative).[10] A degree of *orness* of 0.3 can be said to correspond to a fuzzy degree of *andness* characterized as "*as many as*

[10] I have written a MatLab code that can be used to compute the individual and social degrees of dominance for any values of the two degrees of *orness*.

possible." This means that the decision makers will only be satisfied if "*as many as possible*" of the four criteria are satisfied. This leads to a weights vector:

$w_{11} = 0.15; w_{12} = 0.30; w_{13} = 0.55$ with a maximum entropy of $E(W_1)$=0.97.

As to the stage of social dominance process, I assume that the decision makers are basically interested in the average aggregate at this stage, that is, with a degree of *orness* = 0.5 and a degree of *andness* = 0.5, which would correspond to a weighting vector $w_{21} = w_{22} = w_{23} = w_{24} = 0.25$ and a maximum entropy of $E(W_2)$=1.38. All contributions are hence equally weighed in the social dominance process. Going to Table 3, we obtain the following tables.[11]

Table 6 Weak Bilateralism

AGGREGATION of {DDR=LLL, VUL=HHH, RPL=VHH, RMT=LLL}

		DDR=LLL	*VUL=HHH*	*RPL=VHH*	*RMT=LLL*	
		ID_i^1	ID_i^2	ID_i^3	ID_i^4	
WBB	$SD_{i=1}$	MPP	VLP	MPP	MPP	LPP
SBB	$SD_{i=2}$	AAS	LDD	AAS	AAS	VLD
MUU	$SD_{i=3}$	MDD	LDD	MDD	MDD	MDD
SCC	$SD_{i=4}$	VHD	LDD	VHD	VHD	VHD
	S_{max}	**WBB: Weak Bilateralism**				

Table 7 Strong Bilateralism

AGGREGATION of {DDR=LMM, VUL=VHH, RPL=HHH, RMT=LMM}

		DDR=LMM	*VUL=VHH*	*RPL=HHH*	*RMT=LMM*	
		ID_i^1	ID_i^2	ID_i^3	ID_i^4	
WBB	$SD_{i=1}$	VLP	MDD	VLP	VLP	VLD
SBB	$SD_{i=2}$	LPP	AAS	LPP	LPP	VLP
MUU	$SD_{i=3}$	MDD	AAS	MDD	MDD	MDD
SCC	$SD_{i=4}$	VHD	AAS	VHD	VHD	VHD
	S_{max}	**SBB: Strong Bilateralism**				

[11] These tables are drawn from a larger table that includes all results for all possible combinations of the independent variables for all possible outcomes. The Table (3 pages) is available from the author upon request.

Table 8 Multilateralism

AGGREGATION of {DDR=HHH, VUL=VHH, RPL=LLL, RMT=HHH}

		DDR=HHH	***VUL=VHH***	***RPL=LMM***	***RMT=HHH***	
		ID_i^1	ID_i^2	ID_i^3	ID_i^4	
WBB	$SD_{i=1}$	HDD	MDD	VHD	VHD	DDD
SBB	$SD_{i=2}$	HDD	AAS	MDD	MDD	DDD
MUU	$SD_{i=3}$	LPP	AAS	LPP	LPP	HDD
SCC	$SD_{i=4}$	VLP	AAS	VLP	VLP	HDD
	S_{max}	**MUU: Multilateralism**				

Table 9 Pluralistic Security Community

AGGREGATION of {DDR=VHH, VUL=VHH, RPL=NNN, RMT=VHH}

		DDR=VHH	***VUL=VHH***	***RPL=NNN***	***RMT=VHH***	
		ID_i^1	ID_i^2	ID_i^3	ID_i^4	
WBB	$SD_{i=1}$	VHD	LDD	VHD	VHD	VHD
SBB	$SD_{i=2}$	HDD	AAS	HDD	HDD	MDD
MUU	$SD_{i=3}$	AAS	AAS	AAS	AAS	AAS
SCC	$SD_{i=4}$	MPP	AAS	MPP	MPP	VLP
	S_{max}	**SCC: Pluralistic Security Community**				

For the sake of illustration, let's consider Table 9. The 3rd,..., 6th columns correspond respectively to the four criteria {*DDR, VUL,RPL,RMT*} . The 3rd,..., 6th rows correspond respectively to the four security options {*WBB,SBB,MUU,SCC*}. Hence, on the first criterion, *DDR=VHH* (very high degree of diffuse reciprocity) the option *MUU* is most preferred for the social degree of preference equal to MPP which according to Table 3 ranks better than *VHD*, *HDD*, or *AAS*. Whereas on the criterion of very high vulnerability (*VUL=VHH*), options *SBB*, *MUU*, and *SCC* are equally preferred to the *WBB* option. As can be seen from the table, *SCC* fares better (that is, *MPP*) than all three other options on the remaining criteria of no reliance on power leverage (*RPL=NNN*) and very high reliance on mutual trust (*RMT=VHH*). Aggregating these individual pieces of linguistic information on preference relations among the four security options (that is, aggregating ID_i^1, ID_i^2, ID_i^3, and ID_i^4) clearly shows that *SCC* is the best security alignment, as shown in the last column of Table 11. That is, VLP is higher than AAS, MDD, or

VHD, according to Table 3. In sum, a security community obtains when the degree of diffuse reciprocity is very high, the level of vulnerability is very high, the level of reliance on power leverage is null, and the level of reliance on mutual trust is very high.

A similar analysis when applied to the case of collective decision in NATO leads to Tables B.1 and B.2 in Appendix B. The best alternative is determined as a function of the two degrees of *orness* ω_1 and ω_2 which are varied between 0 and 1. Sanjian's (1992) fuzzy approach predicts that Germany's choice is the best alternative (which his reading of the historical event confirms). A Linguistic fuzzy approach also predicts that the German choice is the best collective alternative for NATO, that is, the option "NEGO." However, as Table B.1 in Appendix B shows there is more to it than just this. Indeed, by taking into account the degrees orness at the individual and social levels all options – POST, NEGO, MONE, and MODE (that is, postpone decision, negotiate with WPA, modernize and negotiate with WPA, and modernize) are possible depending on different degrees of orness ω_1 and ω_2. Moreover, these options are preferred to different levels. For example, as shown in Table B.2 in Appendix B, the option MONE is preferred to a HDD degree for $\omega_2 = 0.05$ and $0.05 \leq \omega_1 \leq 0.35$ and to MDD degree for $\omega_2 = 0.95$ and $0.05 \leq \omega_1 \leq 0.35$.

3 Linguistic Fuzzy-Logic and Social Choice Theory

The LFLA can also contribute to set the ground for a linguistic fuzzy-logic form of social choice theory. Substituting groups for criteria in the formulation of the previous decision problem one can transform practically all decision making problems under multiple criteria into social choice problems, and vice versa. The similarity between aggregation procedures in social choice theory and decision making under multiple criteria is well-known (Nurmi and Meskanen, 2000). For example, the alternatives play the role of the candidates, the criteria play the role of the voters and the decision-maker plays the role of the society.[12] A core problem in this literature pertains to aggregating opinions of individuals to form a collective opinion.

[12] Since Arrow proved his impossibility theorem about half a century ago, many researchers have attempted to avoid his negative result by relaxing some of his original assumptions. In Arrow's framework, a society of k individuals seeks to decide among a set X of discrete alternatives with no additional mathematical structure. Each individual actor has a preference over X which consists of a complete, reflexive, and transitive binary relation over X. A social choice function gives a rule for aggregating any particular profile of individual preferences into a group preference. Allowing preferences of individuals and society or just those of society alone to be fuzzy shows that Arrow's result can, under certain conditions, be avoided using fuzzy preferences and employing a particularly weak version of transitivity among the many plausible definitions of transitivity that are available for fuzzy preferences (e.g., Dutta, 1987), or reaffirmed if strong connectedness is assumed for a variety of weak preference factorizations, even if the transitivity condition is weakened to its absolute minimum (e.g., Richardson, 1998), or with conditionally mixed results (Geslin et al., 2003).

The LFLA opens up the possibility of paying more attention to imprecise preference relations and presents aggregation tools (based on LOWA and MAXENT) that can help explore various possibilities of social choice. Voting is a typical example of collective decision making and hence a natural context for discussing social choice methods. The comparative study of voting procedures has produced a long list of properties or performance criteria that allow for systematic assessments of various voting procedures. These procedures are usually set to satisfy some criteria. The comparison of various procedures of aggregating the votes are evaluated by considering the choices allowed or prohibited by various criteria. The literature on voting is full of various paradoxes that emerge due to various incompatibilities of intuitively plausible requirements regarding social choices. As an illustration consider the situation of four voters VONE, VTWO, VTHREE, and VFOUR who are about to cast their votes on four candidates CAND1, CAND2, CAND3, and CAND4. In conventional social choice theory, we can for example assume that the voters' preference rankings of the candidates from best to worst are as follows (Table 10):

Table 10 Preference rankings from best to worst by voters on the four candidates

Voter	***Preference Ranking of Candidates***
VONE	CAND1 > CAND2 > CAND3 > CAND4
VTWO	CAND2 > CAND3 > CAND4 > CAND1
VTHREE	CAND3 > CAND1 > CAND4 > CAND2
VFOUR	CAND4 > CAND1 > CAND2 > CAND3

For the sake of comparing with the LFLA let us first solve this voting problem using the Borda count procedure. Borda's rule asks voters to rank all the options which are then weighted before aggregation. The higher an individual ranks a particular option, the more points that option will receive. An option is awarded one point for each competitor ranked below it in the voter's ranking. The option with the most total points is declared the winner. Hence, we obtain the following Borda counts:

$$B^{VONE} = \{3,2,1,0\};\ B^{VTWO} = \{0,3,2,1\};\ B^{VTHREE} = \{2,0,3,1\};\ B^{VFOUR} = \{2,1,0,3\}$$

These counts can alternatively be written in a matrix form in the following way (these are the "crisp" analogous of the linguistic ones given below):

$$MB^{VONE} = \begin{bmatrix} - & 1 & 1 & 1 \\ 0 & - & 1 & 1 \\ 0 & 0 & - & 1 \\ 0 & 0 & 0 & - \end{bmatrix} \quad MB^{VTWO} = \begin{bmatrix} - & 0 & 0 & 0 \\ 1 & - & 1 & 1 \\ 1 & 0 & - & 1 \\ 1 & 0 & 0 & - \end{bmatrix} \quad MB^{VTHREE} = \begin{bmatrix} - & 1 & 0 & 1 \\ 0 & - & 0 & 0 \\ 1 & 1 & - & 1 \\ 0 & 1 & 0 & - \end{bmatrix} \quad MB^{FOUR} = \begin{bmatrix} - & 1 & 1 & 0 \\ 0 & - & 1 & 0 \\ 0 & 0 & - & 0 \\ 1 & 1 & 1 & - \end{bmatrix}$$

The Borda count for an option is obtained by adding the elements of a raw for each voter. The total number of points assigned to any option is obtained by summing up the individual ones, which leads to

$$Q^{CAND1} = 7\,;\; Q^{CAND2} = 6\,;\; Q^{CAND3} = 6\,;\; Q^{CAND4} = 5$$

The deterministic Borda count procedure can be turned into a probabilistic Borda count by assigning to every option a probability defined as the Borda score relative to all other options using the average rule:

$$\pi^i = \frac{Q^i}{\sum_j Q^j}$$

We thus obtain:

$$\pi^{CAND1} = \tfrac{7}{24} = 29.2\%\,;\; \pi^{CAND2} = \tfrac{1}{4} = 25\%\,;\; \pi^{CAND3} = \tfrac{1}{4} = 25\%\,;\; \pi^{CAND4} = \tfrac{5}{24} = 20.8\%$$

The actual winner is chosen using a random process based on these weighted probabilities.

Let us now solve the same voting problem using the LFLA approach. To this end, assume that the preference rankings are fuzzy and expressed linguistically. In addition to giving a preference raking of the four candidates for each of the four voters, LFLA allows us to express the preference relations of each voter as a matrix of preference relations expressed symbolically. In order to do this we need to adopt a linguistic terms set for describing the voters' preferences on the candidates. Let's use the semantic set shown in Table 11.

Table 11 Semantic Set for Preference Relations for expressing voters' preferences

k	*Label*	*Linguistic Meaning*
12	*DPP*	*Candidate i* is preferred to *Candidate j* ***in Definite degree***
11	*VHP*	*Candidate i* is preferred to *Candidate j* in ***Very High degree***
10	*HPP*	*Candidate i* is preferred to *Candidate j* in ***High degree***
9	*MPP*	*Candidate i* is preferred to *Candidate j* in ***Moderate degree***
8	*LPP*	*Candidate i* is preferred to *Candidate j* in ***Low degree***
7	*VLP*	*Candidate i* is preferred to *Candidate j* in ***Very Low degree***
6	*AAS*	*Candidate i* is about the same ***AS*** *Candidate j*
5	*VLD*	*Candidate j* is preferred to *Candidate i* in ***Very Low degree***
4	*LDD*	*Candidate j* is preferred to *Candidate i* in ***Low degree***
3	*MDD*	*Candidate j* is preferred to *Candidate i* in ***Moderate degree***
2	*HDD*	*Candidate j* is preferred to *Candidate i* in ***High degree***
1	*VHD*	*Candidate j* is preferred to *Candidate i* in ***Very High degree***
0	*DDD*	*Candidate j* is preferred to *Candidate i* in ***Definite degree***

Using this set we can rewrite the preference relations (binary comparisons of the candidates) as:

$$P^{VONE} = \begin{bmatrix} - & LPP & MPP & HPP \\ LDD & - & LPP & MPP \\ MDD & LDD & - & LPP \\ HDD & MDD & LDD & - \end{bmatrix} \qquad P^{VTWO} = \begin{bmatrix} - & HDD & MDD & LDD \\ HPP & - & LPP & MPP \\ MPP & LDD & - & LPP \\ LPP & MDD & LDD & - \end{bmatrix}$$

$$P^{VTHREE} = \begin{bmatrix} - & MPP & LDD & LPP \\ MDD & - & HDD & LDD \\ LPP & HPP & - & MPP \\ LDD & LPP & MDD & - \end{bmatrix} \qquad P^{VFOUR} = \begin{bmatrix} - & LPP & MPP & LDD \\ LDD & - & LPP & MDD \\ MDD & LDD & - & HDD \\ LPP & MPP & HDD & - \end{bmatrix}$$

Carrying out the same analysis as in the case of NATO choice leads to Table 12 of results depending on the degrees ω_1 and ω_2 of *orness* respectively at the individual and social levels.

Table 12 Social Ranking of the Four Candidates

ω_1	ω_2	CAND1	CAND2	CAND3	CAND4
0.1	0.1	MDD	HDD	HDD	VHD
0.2	0.1	MDD	HDD	HDD	VHD
0.3	0.3	VLD	LDD	LDD	HDD
0.3	0.6	AAS	VLD	VLD	MDD
0.4	0.1	LDD	MDD	MDD	HDD
0.5	0.1	LDD	MDD	MDD	HDD
0.6	0.1	LDD	MDD	MDD	HDD
0.7	0.1	LDD	MDD	MDD	HDD
0.8	0.1	LDD	MDD	MDD	HDD
0.9	0.1	LDD	MDD	MDD	HDD

Table 12 is taken from a larger table which lists all possibilities for the degrees of *orness* at the individual and social levels extending from 0.1 to 0.9.[13] In all other cases not shown in the Table 12 there is no clear winner with CAND1, CAND2, and CAND3 ranked at the same level of preference in all cases. We however see from Table 12 that CAND1 is always a winner, CAND4 is always a looser, while CAND2 and CAND3 are always at the second level. Note that the

[13] The larger table is available from the author upon request.

level of support for the winner (CAND1) generally increases with the degree of *orness* ω_1 from MDD to LDD. There are two exceptions, though, where for $\omega_1 = 0.3$ and $\omega_2 = 0.6$ CAND1 is preferred to a degree AS which is greater than both MDD and LDD. Note that CAND1 is a winner for only $\omega_2 = 0.1$ except for the two values $\omega_2 = 0.3$ and $\omega_2 = 0.6$. In other words, there is a clear winner for various degrees of *orness* at the social level, except again for the two values of $\omega_2 = 0.3$ and $\omega_2 = 0.6$. Whether this is idiosyncratic or a manifestation of a more general result is not clear to me at the moment, but is a question worthy of further exploration. In short, one can arguably state that the LFLA method can reproduce the results obtained through the literature on probabilistic voting in social choice theory (at least in the example considered in this chapter). I can thus venture to argue that the LFLA provides at least another method to study of voting and social choice in addition to the existing rich literature. However, in order to establish the full extent of this claim an axiomatic approach needs to be pursued based on fuzzy (multi-valued) logic and not on classical Boolean (two-valued) logic. In this respect, it is worthwhile emphasizing that LFLA is more than just providing a new way of aggregating linguistic preferences for it is based on a non-Boolean logic the consequences of which are yet to be thoroughly explored for the resolution of lingering dilemmas and paradoxes of social choice theory.

Appendix A

DDR=LLL, $P^{11} = \begin{bmatrix} - & LPP & HPP & DPP \\ LDD & - & MPP & HPP \\ HDD & MDD & - & LPP \\ DDD & HDD & LDD & - \end{bmatrix}$; DDR=LMM, $P^{12} = \begin{bmatrix} - & LDD & HPP & DPP \\ LPP & - & MPP & HPP \\ HDD & MDD & - & LPP \\ DDD & HDD & LDD & - \end{bmatrix}$;

DDR=HHH, $P^{13} = \begin{bmatrix} - & LPP & HDD & DDD \\ LDD & - & MDD & HDD \\ HPP & MPP & - & LPP \\ DPP & HPP & LDD & - \end{bmatrix}$; DDR=VHH, $P^{14} = \begin{bmatrix} - & LDD & HDD & DDD \\ LPP & - & MDD & HDD \\ HPP & MPP & - & LDD \\ DPP & HPP & LPP & - \end{bmatrix}$;

VUL=HHH, $P^{21} = \begin{bmatrix} - & LPP & LPP & LPP \\ LDD & - & AAS & AAS \\ LDD & AAS & - & AAS \\ LDD & AAS & AAS & - \end{bmatrix}$; VUL=VHH, $P^{22} = \begin{bmatrix} - & LDD & LDD & LDD \\ LPP & - & AAS & AAS \\ LPP & AAS & - & AAS \\ LPP & AAS & AAS & - \end{bmatrix}$;

RPL=NNN, $P^{31} = \begin{bmatrix} - & LDD & HDD & DDD \\ LPP & - & MDD & HDD \\ HPP & MPP & - & LDD \\ DPP & HPP & LPP & - \end{bmatrix}$; RPL=LLL, $P^{32} = \begin{bmatrix} - & LDD & HDD & DDD \\ LPP & - & MDD & HDD \\ HPP & MPP & - & LPP \\ DPP & HPP & LDD & - \end{bmatrix}$;

RPL=HHH, $P^{33} = \begin{bmatrix} - & LDD & HPP & DPP \\ LPP & - & MPP & HPP \\ HDD & MDD & - & LPP \\ DDD & HDD & LDD & - \end{bmatrix}$; RPL=VHH, $P^{34} = \begin{bmatrix} - & LPP & HPP & DPP \\ LDD & - & MPP & HPP \\ HDD & MDD & - & LPP \\ DDD & HDD & LDD & - \end{bmatrix}$;

RMT=LLL, $P^{41} = \begin{bmatrix} - & LPP & HPP & DPP \\ LDD & - & MPP & HPP \\ HDD & MDD & - & LPP \\ DDD & HDD & LDD & - \end{bmatrix}$; RMT=LMM, $P^{42} = \begin{bmatrix} - & LDD & HPP & DPP \\ LPP & - & MPP & HPP \\ HDD & MDD & - & LPP \\ DDD & HDD & LDD & - \end{bmatrix}$;

RMT=HHH, $P^{43} = \begin{bmatrix} - & LDD & HDD & DDD \\ LPP & - & MDD & HDD \\ HPP & MPP & - & LPP \\ DPP & HPP & LDD & - \end{bmatrix}$; RMT=VHH, $P^{44} = \begin{bmatrix} - & LDD & HDD & DDD \\ LPP & - & MDD & HDD \\ HPP & MPP & - & LDD \\ DPP & HPP & LPP & - \end{bmatrix}$;

Appendix B

Table B.1 A Fuzzy Set Model of NATO Decision-Making: The Case of Short-Range Nuclear Forces in Europe. Best Collective Alternative.

ω_2 \ ω_1	0.05	0.10	0.15	0.20	0.25	0.30	0.35	0.40	0.45	0.50	0.55	0.60	0.65	0.70	0.75	0.80	0.85	0.90	0.95
0.05	MONE	MONE	MONE	MONE	MONE	MONE	MONE	NEGO	NEGO	NEGO	NEGO	NEGO	NEGO	NEGO	NEGO	NEGO	NEGO	NEGO	NEGO
0.10	POST	POST	POST	POST	POST	POST	POST	POST	POST	NEGO	NEGO	NEGO	MODE	NEGO	MODE	NEGO	MODE	MODE	MODE
0.15	POST	POST	POST	POST	POST	POST	POST	POST	POST	NEGO	NEGO	NEGO	MODE	NEGO	MODE	NEGO	MODE	MODE	MODE
0.20	POST	POST	POST	POST	POST	POST	POST	POST	POST	NEGO	NEGO	NEGO	MODE	NEGO	MODE	NEGO	MODE	MODE	MODE
0.25	POST	POST	POST	POST	POST	POST	POST	NEGO	NEGO	NEGO	NEGO	NEGO	NEGO	NEGO	NEGO	NEGO	NEGO	MODE	MODE
0.30	POST	POST	POST	POST	POST	POST	POST	NEGO	NEGO	NEGO	NEGO	NEGO	NEGO	NEGO	NEGO	NEGO	NEGO	MODE	MODE
0.35	POST	POST	POST	POST	POST	POST	POST	POST	POST	NEGO	NEGO	NEGO	MODE	NEGO	MODE	NEGO	MODE	MODE	MODE
0.40	POST	POST	POST	POST	POST	POST	POST	POST	POST	NEGO	NEGO	NEGO	MODE	NEGO	MODE	NEGO	MODE	MODE	MODE
0.45	POST	POST	POST	POST	POST	POST	POST	POST	POST	NEGO	NEGO	NEGO	MODE	NEGO	MODE	NEGO	MODE	MODE	MODE
0.50	POST	POST	POST	POST	POST	POST	POST	POST	POST	NEGO	NEGO	NEGO	MODE	NEGO	MODE	NEGO	MODE	MODE	MODE
0.55	POST	POST	POST	POST	POST	POST	POST	POST	POST	NEGO	NEGO	NEGO	MODE	NEGO	MODE	NEGO	MODE	MODE	MODE
0.60	POST	POST	POST	POST	POST	POST	POST	POST	POST	NEGO	NEGO	NEGO	MODE	NEGO	MODE	NEGO	MODE	MODE	MODE
0.65	POST	POST	POST	POST	POST	POST	POST	POST	POST	NEGO	NEGO	NEGO	MODE	NEGO	MODE	NEGO	MODE	MODE	MODE
0.70	POST	POST	POST	POST	POST	POST	POST	POST	POST	NEGO	POST	NEGO	MODE	NEGO	MODE	NEGO	MODE	MODE	MODE
0.75	POST	POST	POST	POST	POST	POST	POST	POST	POST	NEGO	POST	NEGO	MODE	NEGO	MODE	NEGO	MODE	MODE	MODE
0.80	POST	POST	POST	POST	POST	POST	POST	POST	POST	NEGO	NEGO	NEGO	MODE	NEGO	MODE	NEGO	MODE	MODE	MODE
0.85	POST	POST	POST	POST	POST	POST	POST	POST	POST	NEGO	NEGO	NEGO	MODE	NEGO	MODE	NEGO	MODE	MODE	MODE
0.90	POST	POST	POST	POST	POST	POST	POST	POST	POST	NEGO	NEGO	NEGO	MODE	NEGO	MODE	NEGO	MODE	MODE	MODE
0.95	MONE	MONE	MONE	MONE	MONE	MONE	MONE	POST	POST	NEGO	NEGO	NEGO	MODE	NEGO	MODE	NEGO	MODE	MODE	MODE

Table B.2 Levels of Preferences for Best Alternative Choice. Level of Preference of Best Collective Alternative.

ω_2 \ ω_1	0.05	0.10	0.15	0.20	0.25	0.30	0.35	0.40	0.45	0.50	0.55	0.60	0.65	0.70	0.75	0.80	0.85	0.90	0.95
0.05	HDD	HDD	HDD	HDD	HDD	HDD	HDD	HDD	HDD	MDD	MDD	LDD	LDD	VLD	VLD	AAS	AAS	AAS	VLP
0.10	MDD	MDD	MDD	MDD	MDD	MDD	MDD	MDD	MDD	MDD	MDD	LDD	LDD	VLD	VLD	AAS	AAS	AAS	VLP
0.15	MDD	MDD	MDD	MDD	MDD	MDD	MDD	MDD	MDD	MDD	MDD	LDD	LDD	VLD	VLD	AAS	AAS	AAS	VLP
0.20	MDD	MDD	MDD	MDD	MDD	MDD	MDD	MDD	MDD	MDD	MDD	LDD	LDD	VLD	VLD	AAS	AAS	AAS	VLP
0.25	MDD	MDD	MDD	MDD	MDD	MDD	MDD	MDD	MDD	LDD	LDD	VLD	VLD	AAS	AAS	VLP	VLP	VLP	LPP
0.30	MDD	MDD	MDD	MDD	MDD	MDD	MDD	MDD	MDD	LDD	LDD	VLD	VLD	AAS	AAS	VLP	VLP	VLP	LPP
0.35	LDD	LDD	LDD	LDD	LDD	LDD	LDD	LDD	LDD	LDD	LDD	VLD	VLD	AAS	AAS	VLP	VLP	VLP	LPP
0.40	LDD	LDD	LDD	LDD	LDD	LDD	LDD	LDD	LDD	LDD	LDD	VLD	VLD	AAS	AAS	VLP	VLP	VLP	LPP
0.45	LDD	LDD	LDD	LDD	LDD	LDD	LDD	LDD	LDD	LDD	LDD	VLD	VLD	AAS	AAS	VLP	VLP	VLP	LPP
0.50	LDD	LDD	LDD	LDD	LDD	LDD	LDD	LDD	LDD	LDD	LDD	VLD	VLD	AAS	AAS	VLP	VLP	VLP	LPP
0.55	LDD	LDD	LDD	LDD	LDD	LDD	LDD	LDD	LDD	LDD	LDD	VLD	VLD	AAS	AAS	VLP	VLP	VLP	LPP
0.60	LDD	LDD	LDD	LDD	LDD	LDD	LDD	LDD	LDD	LDD	LDD	VLD	VLD	AAS	AAS	VLP	VLP	VLP	LPP
0.65	LDD	LDD	LDD	LDD	LDD	LDD	LDD	LDD	LDD	LDD	LDD	VLD	VLD	AAS	AAS	VLP	VLP	VLP	LPP
0.70	LDD	LDD	LDD	LDD	LDD	LDD	LDD	LDD	LDD	LDD	VLD	VLD	VLD	AAS	AAS	VLP	VLP	VLP	LPP
0.75	LDD	LDD	LDD	LDD	LDD	LDD	LDD	LDD	LDD	LDD	VLD	VLD	VLD	AAS	AAS	VLP	VLP	VLP	LPP
0.80	LDD	LDD	LDD	LDD	LDD	LDD	LDD	LDD	LDD	LDD	LDD	VLD	VLD	AAS	AAS	VLP	VLP	VLP	LPP
0.85	LDD	LDD	LDD	LDD	LDD	LDD	LDD	LDD	LDD	LDD	LDD	VLD	VLD	AAS	AAS	VLP	VLP	VLP	LPP
0.90	LDD	LDD	LDD	LDD	LDD	LDD	LDD	LDD	LDD	LDD	LDD	VLD	VLD	AAS	AAS	VLP	VLP	VLP	LPP
0.95	MDD	MDD	MDD	MDD	MDD	MDD	MDD	LDD	LDD	LDD	LDD	VLD	VLD	AAS	AAS	VLP	VLP	VLP	LPP

Chapter 4
Linguistic Fuzzy-Logic 2x2 Games

Conventional game theory faces a dilemma rooted in an incompatibility between the Boolean-logical foundation of the logic of the game and the linguistic nature of strategic communication – linguistic fuzzy logic which "computes with words" all the way down offers a way out of this dilemma. On the one hand, game-theoretic methodologies are inescapably based on a posited underlying logic. The latter is what guides us in judging whether a game-theoretic methodology (and theory) possesses logical coherence and consistency. Boolean two-valued logic underpins conventional game theory. All game-theoretic arguments and the very notion of consistency and inconsistency are based on a sharp Boolean-logic true/false dichotomy. On the other hand, vagueness and equivocation are constitutive features of human communication. Human beings cannot live and communicate without a "language," with the latter being inherently vague. Game theory is essentially the study of strategic communication of information through language in a rigorous and stylized way. However, stylization and rigor cannot totally eradicate vagueness – there is always a remainder of linguistic vagueness at the very heart of the conceptual tools – communicative devices – used in game theory. This creates a foundational dilemma for conventional game theory. Is it possible to resolve this incompatibility between, on the one hand, taking *the dichotomous Boolean two-valued logic* as the logical foundation of game theory and, on the other hand, *the vagueness that inheres in strategic communication* due to the very nature of language? This difficulty is often glossed over or not even sensed at all in conventional game-theoretic works. I propose a remedy to this dilemma by anchoring game theory in linguistic fuzzy logic.

I lay the grounds for and formulate a game-theoretic approach founded not on a Boolean logic, but rather on linguistic fuzzy logic. I thence inscribe vagueness into the logical foundations of game theory. This is a fundamental departure from conventional game theory on two fronts at least. First, positing a fuzzy logic at the foundations of strategic communication and action means that truth values are not only TRUE and FALSE (as in Boolean logic), but can also assume in-between values, such as *very true*, *less true*, *extremely false*, *approximately false*. A number of works have recently been published mostly in the economics literature presenting a fuzzy version of game theory by adopting a fuzzy-set framework, which is conventionally developed using the notion of gradual membership to sets.[1] This work differs from these various efforts on an important aspect, which is the second

[1] Butnariu, 1978; Borges, 1997; Billot, 1992; Maeda, 2000; 2003; Kim and Lee, 2001; Li, 2001; De Wilde, 2004.

B. Arfi: Linguistic Fuzzy Logic Methods in Social Sciences, STUDFUZZ 253, pp. 63–103.
springerlink.com

fundamental departure from conventional game theoretic. I use a linguistic formalism which is anchored in linguistic fuzzy-logic, not fuzzy-set theory. Hence, I do not use the notion of a gradual membership to express the fuzziness of concepts and objects of study. *I "compute with words" all the way down into the inner logic of game theory.*

In order to do this I go back to the logical foundations of game theory and work my way up to strategic interaction. This paves the way for introducing new notions of fuzzy strategy, fuzzy dominant strategy, and fuzzy Nash equilibrium, which are in some sense (i.e., fuzzy-logic sense) extensions of those used in conventional game theory with ordinal preferences. A key concept of this reconceptualization is the notion of truth value of a choice termed in this chapter as the *feasibility value*. Everything becomes fuzzy, everything becomes more or less this or that – everything acquires a linguistic value which is essentially *more or less nuanced and more or less feasible*! We can, for example, talk of nuanced cooperation such as *high cooperation* with a *very high feasibility*, or nuanced cooperation such as *very high cooperation* with a *very low feasibility*, not just of cooperation or no cooperation. If the linguistic fuzzy relations and all other components of the game structure are de-fuzzified by anchoring them in a crisp two-valued logic, the linguistic fuzzy game simply reduces to the conventional game. I apply the new approach to 2x2 Prisoner's Dilemma (PD) game. I find that there is always a strong Nash equilibrium which is Pareto optimal, thereby lifting the dilemma that emerges in the crisp PD game. I begin by analyzing a conventional 2x2 PD game, highlighting the elements of Boolean logic that anchor the logical structure of the game. I then develop the different elements of a linguistic fuzzy-logic game theory – termed as LFL-Game theory. I also formulate a game-theoretic approach to study trust games anchored in linguistic fuzzy logic. I find that there is always an optimum strong Nash equilibrium which is Pareto optimal, thereby lifting many of the dilemmas that emerge in crisp game theory in 2-player trust game. The next chapter extends the ideas of this chapter to the study of social games of cooperation.

1 Boolean Logic and Game Theory

Before introducing key notions of LFL game theory, let us first reformulate a conventional PD game, explicitly using Boolean logic formalism. In the crisp 2x2 PD game the payoffs are defined as: (C,C)=(α, α), (C,D)= (δ, γ), (D,C)= (γ, δ), (D,D)= (β, β), with the condition $\gamma > \alpha > \beta > \delta$. Let s_A and s_B be the crisp choices of A and B, respectively.

Let u_A (u_B) be the crisp payoff of A (B). To clearly define the notion of strategy in linguistic fuzzy logic, from now on I differentiate between initial choices and strategies or strategic arrangements. A set of initial choices is the set of individual moves that an actor might potentially turn into strategies of the game since strategies are strictly defined in games. This differentiation, which is usually

		B	
		C	D
A	C	(α, α)	(δ, γ)
	D	(γ, δ)	(β, β)

Fig. 1 Prisoner's Dilemma Game

redundant in game-theoretic literature, is important because it facilitates the formulation of logical formulas for strategic interaction. Using the concept of inference rules highlights the importance of making such a differentiation, as explained in the following. The rules of inference $\Re_{ij}$ in the game for A (and similar expression for B) in the crisp PD game are:

$\Re_{11}$ If $\{ s_A = C$ and $s_B = C \}$ then $\{ u_A = \alpha \}$ $\Leftrightarrow$ $((s_A = C) \wedge (s_B = C)) \rightarrow (u_A = \alpha)$
$\Re_{21}$ If $\{ s_A = D$ and $s_B = C \}$ then $\{ u_A = \gamma \}$ $\Leftrightarrow$ $((s_A = D) \wedge (s_B = C)) \rightarrow (u_A = \gamma)$
$\Re_{12}$ If $\{ s_A = C$ and $s_B = D \}$ then $\{ u_A = \delta \}$ $\Leftrightarrow$ $((s_A = C) \wedge (s_B = D)) \rightarrow (u_A = \delta)$
$\Re_{22}$ If $\{ s_A = D$ and $s_B = D \}$ then $\{ u_A = \beta \}$ $\Leftrightarrow$ $((s_A = D) \wedge (s_B = D)) \rightarrow (u_A = \beta)$

These rules of inference are based on two-valued Boolean logic. For example, Rule 1 for A translates in English as: **IF** we know that $s_A = C$ is feasible **AND** we know that $s_B = C$ is feasible, **THEN this implies** that $u_A = \alpha$ is feasible. A strategy for a player is then a disjunction of two rules. For example, strategy C for A means that A uses either $\Re_{11}$ or $\Re_{12}$. We thus have for the strategic game as perceived by A (and similar expressions for B):

C: $\Re_{11} \vee \Re_{12}$ $\{((s_A = C) \wedge (s_B = C)) \rightarrow (u_A = \alpha)\} \vee \{((s_A = C) \wedge (s_B = D)) \rightarrow (u_A = \delta)\}$
D: $\Re_{21} \vee \Re_{22}$ $\{((s_A = D) \wedge (s_B = C)) \rightarrow (u_A = \gamma)\} \vee \{((s_A = D) \wedge (s_B = D)) \rightarrow (u_A = \beta)\}$

We can say that A (B) has four initial choices which correspond to the four entries of the normal form of the PD game. These four choices correspond to four inference rules. Therefore, in this notation, a strategy is effectively a disjunction of a number of inference rules. We see, for example, that A (B) has a strictly dominant strategy: D expressed as $\left\{\Re_{21} \vee \Re_{22}\right\}_{A(B)}$ and a strictly dominated strategy: C expressed as $\left\{\Re_{11} \vee \Re_{12}\right\}_{A(B)}$. The solution of the game, which in this case is DD, is the conjunction of the two dominant strategies of the two players:

$$DD \Leftrightarrow \left\{\Re_{21} \vee \Re_{22}\right\}_A \wedge \left\{\Re_{21} \vee \Re_{22}\right\}_B$$

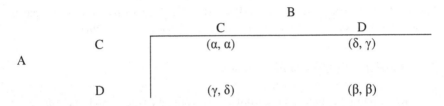

This method of explicating the underlying Boolean logic of the crisp game paves the way for developing a linguistic fuzzy logic approach to game theory.

2 Linguistic Fuzzy-Logic Game

To obtain a linguistic fuzzy game from a crisp game we hence need to change:

1. The initial set of crisp choices of each player into linguistic fuzzy choices: the choices become linguistic variables assuming linguistic values – termed as nuanced choices, each having a feasibility degree.
2. The crisp orderings of the alternative strategies into linguistic orderings: the ranking preferences become linguistic variables assuming linguistic values – termed as nuanced preferences, each with a feasibility degree.
3. The two-valued Boolean logic conjunction AND into linguistic fuzzy logic conjunction, LWC.
4. The two-valued Boolean logic disjunction OR into linguistic fuzzy logic conjunction, LWD.
5. The two-valued Boolean logic implication → into linguistic fuzzy logic implication, LI.
6. The rules of the game $\mathfrak{R}_{ij}$: changing them from inference based on two-valued logic to inference based on linguistic fuzzy logic.

Note that to each choice we hence associate a primary term, a linguistic degree of nuance expressed using a linguistic hedge ν applied to the primary term of the variable, and a feasibility value expressed using a linguistic hedge φ applied to the primary term of a designated feasibility value such as *feasible*. For example, for a feasible nuanced choice of cooperation we would have the term set of hedged ν-cooperation. We could, for example, have *low-cooperation*, *moderate-cooperation*, *high-cooperation*, etc. The hedges (or *nuancers*) in this case are $\nu \in$ {*low*, *moderate*, *high*}. Each of these linguistic values of nuanced cooperation would have a linguistic degree of feasibility such as, for example, *low-to-moderate feasibility*, *high-feasibility*, and *moderate-to-high feasibility*, respectively; hence we have the set of feasibility hedges $\varphi \in$ {*low-to-moderate*, *high*, *moderate-to-high*}. The player's choice set would be represented as $s = \{(\nu\text{-}cooperation, \varphi\text{-}feasible), \text{etc.}\}$. Formally, we would write: $s_i = \left(\nu_{[s_i=\ldots]}, \varphi_{[s_i=\ldots]}\right)$. The degree of feasibility can be interpreted in a PD game as a player's predisposition toward being more nice-spirited or more mean-spirited (as illustrated down below in a full solution of PD game). The LFL approach hence allows us to explore the impact of players' predispositions as part of the game structure – *players' predispositions are endogenous in the LFL-game*, rather than being assumed as types given by nature, as is usually done in conventional game theory. This opens up the possibility for richer notions of strategic dominance and Nash equilibrium, as discussed shortly. Similarly, we represent the preference rankings as $R_i = \left(v_{[R_i=...]}, \varphi_{[R_i=...]}\right)$ and the inference rules as $\Re_{ij} = \left(v_{[\Re_{ij}=...]}; \varphi_{[\Re_{ij}=...]}\right)$.

The next step is to express the rules of the game in a linguistic fuzzy logic form. This would enable us to define the notions of a linguistic fuzzy-logic dominant strategy and linguistic fuzzy-logic Nash equilibrium. In doing so we must keep in mind that all variables are linguistically expressed and symbolically manipulated, as well as all convex combinations thereof formed by using the four connectives ($\wedge$ = AND, $\vee$ = OR, $\rightarrow$ = Implication, $\neg$ = Negation). To this end, I use the algebra introduced earlier in the book.

To fuzzify a PD game, we start with the ordinal preferences of an ordinal PD game. Each player has a matrix of preference relations for a particular strategy. In the case of the crisp 2x2 PD game A's preference relation would hence be $R_A(C;C)$ for both A and B cooperating and $R_A(C;D)$ for A cooperating while B defecting. A's and B's respective matrices of preference relations are given as:

$$\text{A: } R_A = \begin{pmatrix} R_A(C;C) & R_A(C;D) \\ R_A(D;C) & R_A(D;D) \end{pmatrix}; \qquad \text{B: } R_B = \begin{pmatrix} R_B(C;C) & R_B(C;D) \\ R_B(D;C) & R_B(D;D) \end{pmatrix}$$

In the case of the PD game we have: $R_A(D;C) > R_A(C;C) > R_A(D;D) > R_A(C;D)$ and $R_B(D;C) > R_B(C;C) > R_B(D;D) > R_B(C;D)$. The various choices are ranked from best to worst as 1,2,3,4. The dilemma in the PD game is that both A and B end up rationally getting the outcome (3,3) by seeking to obtain either (1,4) for A or (4,1) for B – the outcome (3,3) corresponding to mutual defection is worse than that corresponding to mutual cooperation (2,2). We can express the crisp game using the rules of inference of the underlying Boolean logic for player A as:

$$\Re_{11} \quad \text{If}\{ s_A = C \text{ and } s_B = C \}\text{then}\{ R_A(C;C) = 2 \} \quad \Leftrightarrow \quad ((s_A = C)\wedge(s_B = C)) \rightarrow (R_A(C;C) = 2)$$
$$\Re_{21} \quad \text{If}\{ s_A = D \text{ and } s_B = C \}\text{then}\{ R_A(D;C) = 1 \} \quad \Leftrightarrow \quad ((s_A = D)\wedge(s_B = C)) \rightarrow (R_A(D;C) = 1)$$
$$\Re_{12} \quad \text{If}\{ s_A = C \text{ and } s_B = D \}\text{then}\{ R_A(C;D) = 4 \} \quad \Leftrightarrow \quad ((s_A = C)\wedge(s_B = D)) \rightarrow (R_A(C;D) = 4)$$
$$\Re_{22} \quad \text{If}\{ s_A = D \text{ and } s_B = D \}\text{then}\{ R_A(D;D) = 3 \} \quad \Leftrightarrow \quad ((s_A = D)\wedge(s_B = D)) \rightarrow (R_A(D;D) = 3)$$

(and similar expressions for B). We can rewrite these by using the notion of strategy as defined earlier in terms of inference rules.

$$\text{C}: \Re_{11} \vee \Re_{12} \quad \{((s_A = C)\wedge(s_B = C)) \rightarrow (R_A(C;C) = 2)\} \vee \{((s_A = C)\wedge(s_B = D)) \rightarrow (R_A(C;D) = 4)\}$$
$$\text{D}: \Re_{21} \vee \Re_{22} \quad \{((s_A = D)\wedge(s_B = C)) \rightarrow (R_A(D;C) = 1)\} \vee \{((s_A = D)\wedge(s_B = D)) \rightarrow (R_A(D;D) = 3)\}$$

It is clear that in both cases D is a strictly dominant strategy for both players. The Nash equilibrium is thus $DD \Leftrightarrow \{\Re_{21} \vee \Re_{22}\}_A \wedge \{\Re_{21} \vee \Re_{22}\}_B$.

To move from a crisp game to a linguistic fuzzy-logic (LFL)-game we transform the above crisp relations into linguistic fuzzy preference relations. Before doing so, let us consider as an example the linguistic fuzzy-logic equivalent of Rule 11 for A: $\Re_{11} = s_A^1 \wedge s_B^1 \rightarrow R_A\left(s_A^1; s_B^1\right)$, that is:

To what degrees of nuance $v_{\Re_{11}}^A$ *and feasibility* $\varphi_{\Re_{11}}^A$ *can we infer that:*

IF [{strategy s_A^1 *has nuance degree* $v_{[s_A^1]}$*} with a feasibility degree* $\varphi_{[s_A^1]}$*]*

AND [{strategy s_B^1 *has nuance degree* $v_{[s_B^1]}$*} with a feasibility degree* $\varphi_{[s_B^1]}$*]*

THEN [{preference relation $R_A\left(s_A^1; s_B^1\right)$ *has nuance degree* $v_{R_A\left(s_A^1;s_B^1\right)}$*} with a feasibility degree* $\varphi_{R_A\left(s_A^1;s_B^1\right)}$*]?*

Using the LWC and LI operations we write the feasibility and nuance degrees of Rule 11 for A:

$$\left(v_{\Re_{11}}^A;\varphi_{\Re_{11}}^A\right) = LI\left\{LWC\left[\left(v_{[s_A]},\varphi_{[s_A]}\right),\left(v_{[s_B]},\varphi_{[s_B]}\right)\right],\left(v_{R_A(s_A;s_B)},\varphi_{R_A(s_A;s_B)}\right)\right\} \tag{39}$$

The linguistic fuzzy logic equivalents of the remaining rules (12; 21; 22) are written in similar fashions. A strategic choice s_A^1 for player A then consists of choosing a linguistic fuzzy-logic strategy as a disjunction of linguistic fuzzy-logic inference rules (formally, just like in a crisp game):

$$\left\{\Re_{11} \vee \Re_{12}\right\}_A = \left\{s_A^1 \wedge s_B^1 \rightarrow R_A\left(s_A^1; s_B^1\right)\right\} \vee \left\{s_A^1 \wedge s_B^2 \rightarrow R_A\left(s_A^1; s_B^2\right)\right\} \tag{40}$$

The degrees of nuance and feasibility for this strategy are given by:

$$\left(v_{11,12}^A,\varphi_{11,12}^A\right) = Disjunction\left\{\left(v_{\Re_{11}}^A,\varphi_{\Re_{11}}^A\right);\left(v_{\Re_{12}}^A;\varphi_{\Re_{12}}^A\right)\right\} = LWD\left\{\Re_{11};\Re_{12}\right\}_A \tag{41}$$

Rationally choosing a strategy then would consist in optimizing the degrees of nuance and feasibility – that is, choosing a strategy with the highest degree of nuance and/or the highest degree of feasibility of the disjunction of rules ($\Re_{11}$ or $\Re_{12}$). An outcome (profile) of the game would then be a linguistic conjunction of two strategies of the two players, e.g., $\left\{\Re_{11} \vee \Re_{12}\right\}_A \wedge \left\{\Re_{11} \vee \Re_{12}\right\}_B$. The degrees of nuance and feasibility of this outcome will be obtained as:

$$\left(v_{GAME}, \varphi_{GAME}\right) = LWC\left\{LWD\left\{\Re_{11};\Re_{12}\right\}_A ; LWD\left\{\Re_{11};\Re_{12}\right\}_B\right\} \quad (42)$$

At this juncture, let us introduce some notations that would pave the way for a generalized definition of a dominant linguistic fuzzy-logic strategy and an LFL-Game profile. Let P be the set of q players. Let h be the total number of hedges (forming a set H) available to the players. If, for example, in the crisp ordinal game each player possesses one initial crisp choice, the player can apply a maximum of h hedges to this initial choice and hence would end up having to choose among h different linguistic fuzzy-logic choices, each having a linguistic degree of nuance and a linguistic degree of feasibility. Hence, for a player i we have a set of possible linguistic fuzzy-logic choices defined as:

$$S_i = \left\{ s_i^{l_i}\left(v_{l_i}, \varphi_{l_i}\right) \middle| l_i = 1,\ldots,q_i ; v_{l_i} \in H ; \varphi_{l_i} \in H \right\} \quad (43)$$

That is: each player i has q_i LFL-choices and each of these choices has its own feasibility value φ and nuance value ν. For example, in the case of the crisp PD game, A has two crisp choices, C and D. From these two crisp choices we would have two sets of LFL-choices obtained by applying the elements of the hedge algebra set H to these crisp choices. Suppose, for example, that the hedge set is given by:

H= {*NNN, VLL, LLL, MMM, HHH, VHH, FFF*}

where:

NNN=Null, VLL=Very Low, LLL=Low, MMM=Moderate, HHH=High, VHH=Very High, FFF=Full

A would have the following two sets of LFL-choices:

{ν-C| ν ∈ *H*}={ *NNN, VLL, LLL, MMM, HHH, VHH, FFF*}*-cooperation*
{ν-D| ν ∈ *H* }={ *NNN, VLL, LLL, MMM, HHH, VHH, FFF*}*-defection*

Each of these LFL-choices would have a feasibility degree:

φ ∈{ *NNN, VLL, LLL, MMM, HHH, VHH, FFF*}*-feasibility*

A could, for example, choose MMM-C (*Moderate-cooperation*) with VHH-(*Very High degree of) feasibility*. The sets {ν-C| ν ∈ *H*} and {ν-D| ν ∈ *H*} are semantically speaking the mirror images of one another; that is, for example, *null-cooperation* is the same as *full-defection*. It is therefore enough to consider the set of LFL-choices defined as {ν-C| ν ∈ *H*}. This is not always the case, especially starting with crisp games where the choice set is not two-dimensional or, more generally, when the crisp choices are not necessarily symmetrically related to one another in terms of their semantics.

If we write the linguistic feasibility values φ and the linguistic nuance degrees v for the q_i choices for player i as:

$$\left\{\varphi_i\left(s_i^k\right) \mid k=1,\ldots,q_i\right\};\left\{\nu_i\left(s_i^k\right) \mid k=1,\ldots,q_i\right\} \quad (44)$$

The linguistic feasibility and nuance values for an inference rule $\Re_{kl}$ for player i would be given by:

$$\left(\nu_{\Re_{kl}}^i;\varphi_{\Re_{kl}}^i\right)=LI\left\{LWC\left\{\left(\nu_i\left(s_i^k\right),\varphi_i\left(s_i^k\right)\right);\left(\nu_i\left(s_{-i}^l\right),\varphi_i\left(s_{-i}^l\right)\right)\right\};\left(\nu_i\left[R_i\left(s_i^k,s_{-i}^l\right)\right],\varphi_i\left[R_i\left(s_i^k,s_{-i}^l\right)\right]\right)\right\} \quad (45)$$

where, following a standard practice, I denote by a_{-i} the *n-1* tuple that results from removing the *i-th* element from the *n*-tuple $a=\left(a_1,\cdots,a_n\right)$. The set of LFL-choices that player *i*'s opponents may play is then denoted by:

$$S_{-i}=S_1\times\cdots\times S_{i-1}\times S_{i+1}\times\cdots S_q \quad (46)$$

The linguistic fuzzy-logic choices of all other players $j \neq i$ are

$$S_{-i}=\left\{s_j^{l_j}\left(\nu_{l_j},\varphi_{l_j}\right)\Big| j=1,2,\ldots,i-1,i+1,\ldots,q;l_j=1,\ldots,q_j;\nu_{l_j}\in H;\varphi_{l_j}\in H\right\} \quad (47)$$

The feasibility and nuance values for a player's choice are functions of three prior feasibility values and three prior degrees of nuance as the following example from the PD game shows:

$$\left(\nu_{\Re_{11}}^A;\varphi_{\Re_{11}}^A\right)=LI\left\{LWC\left[\left(\nu_{[MMM]},\varphi_{[MMM]}\right),\left(\nu_{[HHH]},\varphi_{[HHH]}\right)\right],\left(\nu_{R_A(MMM;HHH)},\varphi_{R_A(MMM;HHH)}\right)\right\} \quad (48)$$

For example, for player A the feasibility value $\varphi_{\Re_{11}}^A$ for playing $\Re_{11}$ is a function of:

- *The degrees of nuance and corresponding feasibility values* $\left(\nu_{[s_A=MMM]},\varphi_{[s_A=MMM]}\right)$ *for A playing MMM-C (moderate-cooperation) with a degree of feasibility* $\varphi_{[s_A=MMM]}$ *and* $\left(\nu_{[s_B=HHH]},\varphi_{[s_B=HHH]}\right)$ *for B playing HHH-C (high-cooperation) with a degree of feasibility* $\varphi_{[s_B=HHH]}$
- *The degree of nuance and corresponding feasibility value of preference relation of player A should this rule to be chosen,* $\left(\nu_{R_A(MMM;HHH)},\varphi_{R_A(MMM;HHH)}\right)$.

The set of inference rules available to i playing against j are thus generally written as:

$$\Re_{l_i\,l_j} = \left\{ s_i^{l_i} \wedge s_j^{l_j} \rightarrow R_i\left(s_i^{l_i}; s_j^{l_j}\right) \middle| l_i \in \{1,\ldots,q_i\}; l_j \in \{1,\ldots,q_j\} \right\} \quad (49)$$

The strategy of player i consists of a disjunction of all j's choices, which in effect is nothing but player i's strategy when player i's initial choice is $s_i^{l_i}$. Hence, we can write the strategy $\zeta(.)$ for player i against player j as:

$$\zeta\left(s_i^{l_i}\right) = \Re_{l_i\,(l_j)} = \Re_{l_i\,k_1} \vee \Re_{l_i\,k_2} \vee \ldots \vee \Re_{l_i\,k_{q_j}} = LWD\left\{\Re_{l_i\,k_1}; \Re_{l_i\,k_2}; \ldots; \Re_{l_i\,k_{q_j}}\right\} \quad (50)$$

The linguistic nuance and feasibility values of i's strategy for when player i's choice is $s_i^{l_i}$ are:

$$\zeta\left(v_i^{l_i}; \varphi_i^{l_i}\right) = LWD\left\{\left(v_{\Re_{l_i k_1}}; \varphi_{\Re_{l_i k_1}}\right); \left(v_{\Re_{l_i k_1}}; \varphi_{\Re_{l_i k_2}}\right); \ldots; \left(v_{\Re_{l_i k_1}}; \varphi_{\Re_{l_i k_{qj}}}\right)\right\} \quad (51)$$

We now are in a position to define three types of strictly dominant strategies.

Strictly F- Dominant Strategy

A strictly F-dominant linguistic fuzzy-logic strategy $\zeta\left(s_i^k\right)$ *with a feasibility value* φ_i^k *for player i is such that* $\varphi_i^k \succ \varphi_i^{[-k]}$ *where [–k] stands for all linguistic fuzzy-logic strategies for player i other than strategy k. The relation of domination is partial if the strict relation* $\succ$ *is replaced by the weaker relation,* $\succsim$.

Strictly N- Dominant Strategy

A strictly N-dominant linguistic fuzzy-logic strategy $\zeta\left(s_i^k\right)$ *with a nuance value* v_i^k *for player i is such that* $v_i^k \succ v_i^{[-k]}$ *where [–k] stands for all linguistic fuzzy-logic strategies for player i other than strategy k. The relation of domination is partial if the strict relation* $\succ$ *is replaced by the weaker relation,* $\succsim$.

Strictly F-N- Dominant Strategy

A strictly F-N-dominant linguistic fuzzy-logic strategy $\zeta\left(s_i^k\right)$ *with a nuance value* v_i^k *and a feasibility value* φ_i^k *for player i is such that* $v_i^k \succ v_i^{[-k]}$ *and*

$\varphi_i^k \succ \varphi_i^{[-k]}$ where [–k] stands for all linguistic fuzzy-logic strategies for player i other than strategy k. The relation of domination is partial if the strict relation $\succ$ is replaced by the weaker relation, $\succsim$.

Since the hedges are always (at least partially) ordered, the degrees of nuance (and the degrees of feasibility) for any two strategies of a player stand (at least partially) ordered toward one another. This implies that a player has always at least one partially, if not strictly, dominant strategy in one of the three meanings defined above. We now can provide the following complete definition.

Definition: 2-Player Linguistic Fuzzy-Logic (LFL) Game:

A 2X2 LFL-Game G is a couple (P,H), where P is a finite set of 2 players and H a finite set of h linguistic hedges, with the following elements:

1. *For each player $i \in P$ there is a finite set of q_i initial LFL-choices*

$$S_i = \left\{ s_i^{l_i}\left(v_{l_i}, \varphi_{l_i}\right) \,\middle|\, l_i = 1,\ldots,q_i;\ v_{l_i} \in H;\ \varphi_{l_i} \in H \right\} \tag{52}$$

   - *$\varphi_{l_i} : S_i \rightarrow H$ is called the feasibility value of the LFL-choice $s_i^{l_i}\left(v_{l_i}, \varphi_{l_i}\right)$*
   - *$v_{l_i} : S_i \rightarrow H$ is called the nuance degree of the LFL-choice $s_i^{l_i}\left(v_{l_i}, \varphi_{l_i}\right)$*

2. *For each player $i \in P$ there is a $R_i\left(s_i^{l_i}; s_{-i}\right)$ matrix of preference relations with $q_1 \times q_2$ elements showing i's subjective ranking of simultaneous choices for both players in the game. Each matrix element has a nuance degree spanning a spectrum from fully preferred (i.e., best) to not preferred (i.e., worst) and a feasibility degree spanning a spectrum from fully feasible to not feasible.*
3. *The LFL-Game G has $q_1 \times q_2$ rules: Rule $\Re_{kl}$ for $i \in P$ playing s_i^k against $j \in P$ playing s_j^l is:*

$$\left\{ \Re_{kl} = \left[\left(s_i^k \wedge s_j^l\right) \rightarrow R_i\left(s_i^k; s_j^l\right) \right] \,\middle|\, k \in \{1,\ldots,q_1\}; l \in \{1,\ldots,q_2\}; i \in \{1,2\} \neq j \in \{1,2\} \right\} \tag{53}$$

The degrees of nuance v and feasibility φ of $\Re_{kl}$ are given by:

$$\left(v_{\mathfrak{R}_{kl}}^{i};\varphi_{\mathfrak{R}_{kl}}^{i}\right)=LI\left\{LWC\left\{\left(v_{i}\left(s_{i}^{k}\right),\varphi_{i}\left(s_{i}^{k}\right)\right);\left(v_{i}\left(s_{j}^{l}\right),\varphi_{i}\left(s_{j}^{l}\right)\right)\right\};\left(v_{i}\left[R_{i}\left(s_{i}^{k},s_{j}^{l}\right)\right],\varphi_{i}\left[R_{i}\left(s_{i}^{k},s_{j}^{l}\right)\right]\right)\right\} \tag{54}$$

4. *The LFL-strategic arrangement for player i against player j when player i's initial choice is* $s_i^{l_i}$ is:

$$\zeta\left(s_i^{l_i}\right)=\mathfrak{R}_{l_i\ (k_j)}=\mathfrak{R}_{l_i\ k_1}\vee\mathfrak{R}_{l_i\ k_2}\vee\ldots\vee\mathfrak{R}_{l_i\ k_{q_j}}=LWD\left\{\mathfrak{R}_{l_i\ k_1};\mathfrak{R}_{l_i\ k_2};\ldots;\mathfrak{R}_{l_i\ k_{q_j}}\right\} \tag{55}$$

The nuance and feasibility values for this strategic arrangement are:

$$\left(v_i^{l_i};\varphi_i^{l_i}\right)=LWD\left\{\left(v_{\mathfrak{R}_{l_ik_1}};\varphi_{\mathfrak{R}_{l_ik_1}}\right);\left(v_{\mathfrak{R}_{l_ik_1}};\varphi_{\mathfrak{R}_{l_ik_2}}\right);\ldots;\left(v_{\mathfrak{R}_{l_ik_1}};\varphi_{\mathfrak{R}_{l_ik_{q_j}}}\right)\right\} \tag{56}$$

5. *The Cartesian product* $S=S_1\times S_2$ *is called the LFL-strategy profile for the LFL-Game G.*
A profile GP for players i and j in the LFL-Game is given by:

$$GP\left(i,l_i;j,n_j\right)=\mathfrak{R}_{l_i(k_j)}\wedge\mathfrak{R}_{(m_i)n_j}=LWC\left\{\mathfrak{R}_{l_i(k_j)};\mathfrak{R}_{(m_i)n_j}\right\} \tag{57}$$

The nuance and feasibility degrees of this profile are obtained as:

$$\left(v_{l_il_j}^{pr};\varphi_{l_il_j}^{pr}\right)=LWC\left\{\zeta\left(v_i^{l_i};\varphi_i^{l_i}\right);\zeta\left(v_j^{l_j};\varphi_j^{l_j}\right)\right\} \tag{58}$$

We can recover the case of a crisp 2x2 Boolean-logic PD game by setting:

P={A,B}

H={NNN, FFF} or equivalently H={false, true}

$$S=\{C;D\}\times\{C;D\}$$

$$R_A(D;C)\succ R_A(C;C)\succ R_A(D;D)\succ R_A(C;D)$$

$$R_B(D;C)\succ R_B(C;C)\succ R_B(D;D)\succ R_B(C;D)$$

To define a notion of Nash equilibrium let us recall that the notion of Nash equilibrium for crisp ordinal game is usually introduced through an ordering operation that ranks the preference relations of the players. In the LFL approach the notion of dominant strategy is defined by using the degrees of nuance and feasibility, which are by construction always ordered. Because an LFL-strategy has both a degree of nuance and a degree of feasibility we can have three sorts of Nash equilibria.

Definition: Nash Equilibrium:

Each player $i \in \{1,2\}$ *has a finite set of linguistic feasibility values* φ_i^l *and a finite set of linguistic nuance values* v_i^l *defining its LFL-strategies:*

$$\left\{\zeta\left(s_i^{l_i}\right)=\zeta\left(v_i^{l_i};\varphi_i^{l_i}\right)\middle| i\in\{1,2\};l_i\in\{1,\ldots,q_i\}\right\} \tag{59}$$

The game profile is given by:

$$\left(v_{l_1l_2}^{pr};\varphi_{l_1l_2}^{pr}\right)=LWC\left\{\zeta\left(v_1^{l_1};\varphi_1^{l_1}\right);\zeta\left(v_2^{l_2};\varphi_2^{l_2}\right)\right\} \tag{60}$$

NNE: N-Nash equilibrium

$\left(v^*{}_{l_1l_2}^{pr};\varphi_{l_1l_2}^{pr}\right)$ *is called an N-Nash equilibrium (NNE) if:*

$$\left(v_{l_1l_2}^{pr}\prec v^*{}_{l_1^*l_2^*}^{pr}\right),\ \forall l_1\in\{1,\ldots,1-l_1^*,1+l_1^*,\ldots,q_1\}\ ;\ \forall l_2\in\{1,\ldots,1-l_2^*,1+l_2^*,\ldots,q_2\} \tag{61}$$

FNE: F-Nash Equilibrium:

$\left(v_{l_1l_2}^{pr};\varphi^*{}_{l_1l_2}^{pr}\right)$ *is called an F-Nash equilibrium (FNE) if:*

$$\left(\varphi_{l_1l_2}^{pr}\prec \varphi^*{}_{l_1^*l_2^*}^{pr}\right),\ \forall l_1\in\{1,\ldots,1-l_1^*,1+l_1^*,\ldots,q_1\}\ ;\ \forall l_2\in\{1,\ldots,1-l_2^*,1+l_2^*,\ldots,q_2\} \tag{62}$$

FNNE: F-N-Nash Equilibrium:

$\left(v^*{}_{l_1l_2}^{pr};\varphi^*{}_{l_1l_2}^{pr}\right)$ *is called an F-N-Nash equilibrium (FNNE) if:*

$$\left(v_{l_1l_2}^{pr}\prec v^*{}_{l_1^*l_2^*}^{pr}\right)\text{ and }\left(\varphi_{l_1l_2}^{pr}\prec \varphi^*{}_{l_1^*l_2^*}^{pr}\right),$$

$$\forall l_1\in\{1,\ldots,1-l_1^*,1+l_1^*,\ldots,q_1\}\ ;\ \forall l_2\in\{1,\ldots,1-l_2^*,1+l_2^*,\ldots,q_2\} \tag{63}$$

Because the set of hedges is an ordered algebra and the operations of LOWA, LWC, LWD, and LI are convex operations, we can see that the processes of linguistic aggregation produce linguistic terms which are elements of the same set of partially ordered hedges. This implies that there will always be a partial ordering of the profiles of a game. That is, the degrees of nuance and feasibility of the various profiles are always partially ordered as elements of the hedge algebra. This in turn implies that that there will at least be an N or an F Nash equilibrium as defined up above. The simultaneous occurrence of F and N types of equilibrium produces an F-N Nash equilibrium. We thus can state the following theorem of existence.

Theorem: Existence of Nash Equilibrium

There always exists at least one Nash equilibrium in an LFL game. The Nash equilibrium can be of F type, N type, or both simultaneously, that is, F-N type.

To illustrate these notions of LFL-Nash equilibrium let us consider the following game structure:
The basic crisp strategy is cooperation.

- The hedge algebra with seven degrees of nuance and feasibility is: {NNN, VLL, LLL, MMM, HHH, VHH, FFF}[2]
- Assume that each player has seven LFL-choices $\left\{ s_i^j \middle| i = 1,2; j = 1,2,3,4,5,6,7 \right\}$ with the following nuance degrees: {NNN, VLL, LLL, MMM, HHH, VHH, FFF}. This means that the choices are: no-cooperation, very-low-cooperation, low-cooperation, moderate-cooperation, high-cooperation, very-high-cooperation, and full-cooperation.
- Assume that player 1 has the following feasibility degrees for its LFL-choices: {LLL, HHH, VHH, VLL, FFF, NNN, LLL}. This means player 1 perceives its choice as being no-cooperation with a low degree of feasibility, very-low-cooperation with a high degree of feasibility, low-cooperation with a very high degree of feasibility, moderate-cooperation with a very low degree of feasibility, high-cooperation with a full degree of feasibility, very-high-cooperation with null degree of feasibility, and full-cooperation with low degree of feasibility
- Assume that player 2 has the following feasibility degrees for its LFL-choices: {HHH, MMM, VHH, FFF, LLL, NNN, VLL}. This means player 2 perceives its choices as being no-cooperation with a high degree of feasibility, very-low-cooperation with a moderate degree of feasibility, low-cooperation with a very high degree of feasibility, moderate-cooperation with a full degree of feasibility, high-cooperation with a low degree of feasibility, very-high-cooperation with null degree of feasibility, and full-cooperation with very low degree of feasibility.

The elements of the matrix of strategic preference relations for player 1 (and similar expression for player 2) are $\left\{ R_1\left(s_1^i; s_2^j\right) \middle| i = 1,2,3,4,5,6,7;\ j = 1,2,3,4,5,6,7 \right\}$. Each of these elements has a nuance degree spanning a spectrum from fully preferred (i.e., best) to not preferred (i.e., worst) and a feasibility degree spanning a spectrum from fully feasible to not feasible. We thus have as input for each player two 7x7 matrices, one for the degrees of nuance and one for the degrees of feasibility. These elements are defining features of the LFL-Game much like the utility entries define crisp cardinal and ordinal games. The corresponding matrix for a crisp PD game would read using the following ordinal utilities (with best = 1, …, worst = 4):

[2] NNN=null, VLL=very low, LLL=low, MMM=moderate, HHH=high, VHH=very high, and FFF=full.

$$R_1^{ordinal} = \begin{array}{c} \\ C \\ D \end{array}\overset{\begin{array}{cc} C & D \end{array}}{\begin{bmatrix} 2 & 4 \\ 1 & 3 \end{bmatrix}}\ ;\ R_2^{ordinal} = \begin{array}{c} \\ C \\ D \end{array}\overset{\begin{array}{cc} C & D \end{array}}{\begin{bmatrix} 2 & 1 \\ 4 & 3 \end{bmatrix}} \tag{64}$$

For an LFL-Game we have the matrix of preference relations for player 1:

$$R_1 = \left\{ R_1\left(s_1^i; s_2^j\right) \middle| i = 1,\ldots,7;\ j = 1,\ldots,7 \right\} \tag{65}$$

Each element of this matrix has a nuance degree and a feasibility degree. What matters for the assessment of a rule and the possibility of having Nash equilibrium are these degrees of nuance and feasibility. To proceed further in the illustration of an LFL-Game we hence need to postulate these values as a further specification of the structure of the LFL-Game. To keep a transparent connection with the Boolean-logic ordinal PD game we need to remember that for the latter we have all degrees of feasibility equal to TRUE (or, in LFL denotation, FFF). As to the degrees of nuance we have for player 1 the following rankings: 1 for strategy DC, 2 for strategy CC, 3 for strategy DD, and 4 for strategy CD. These rankings translate into the following degrees of nuance in LFL terminology: FFF for DC, HHH for CC, LLL for DD, and NNN for CD. We thus have the following two matrices for both players:

Matrix of Preference Nuances in crisp ordinal PD game

$$\textit{Player 1: } \nu\left[R_1^{ordinal}\right] = \begin{bmatrix} HHH & NNN \\ FFF & LLL \end{bmatrix};\ \textit{Player 2: } \nu\left[R_2^{ordinal}\right] = \begin{bmatrix} HHH & FFF \\ NNN & LLL \end{bmatrix} \tag{66}$$

Matrix of Preference Feasibilities in crisp ordinal PD game

$$\textit{Player 1: } \varphi\left[R_1^{ordinal}\right] = \begin{bmatrix} FFF & FFF \\ FFF & FFF \end{bmatrix};\ \textit{Player 2: } \varphi\left[R_2^{ordinal}\right] = \begin{bmatrix} FFF & FFF \\ FFF & FFF \end{bmatrix} \tag{67}$$

In the LFL-PD game we would, for example, write the matrix of nuanced preferences as:

Matrix of nuanced preferences:

$$\nu\left[R_1\right] = \begin{bmatrix} HHH & MMM & LLL & VLL & NNN & NNN & NNN \\ HHH & MMM & LLL & VLL & NNN & NNN & NNN \\ VHH & HHH & MMM & LLL & VLL & NNN & NNN \\ VHH & HHH & MMM & LLL & VLL & VLL & VLL \\ VHH & HHH & MMM & LLL & VLL & VLL & VLL \\ FFF & VHH & HHH & MMM & LLL & VLL & VLL \\ FFF & VHH & HHH & MMM & LLL & VLL & VLL \end{bmatrix} \tag{68}$$

Matrix of degrees of feasibility:

$$\varphi[R_1] = \begin{bmatrix} FFF & VHH & HHH & MMM & HHH & VHH & FFF \\ VHH & HHH & MMM & LLL & MMM & HHH & VHH \\ HHH & MMM & LLL & VLL & LLL & MMM & HHH \\ MMM & LLL & VLL & NNN & VLL & LLL & MMM \\ HHH & MMM & LLL & VLL & LLL & MMM & HHH \\ VHH & HHH & MMM & LLL & MMM & HHH & VHH \\ FFF & VHH & HHH & MMM & HHH & VHH & FFF \end{bmatrix} \tag{69}$$

This means, for example, that for player 1 $\nu\left[R_1\left(s_1^3;s_2^5\right)\right]=VLL$, that is, player 1 ranks a strategy with (very-high-cooperation, no-cooperation) (3rd row, 5th column) as having a very low degree of nuance of preference ranking. We also need a matrix of feasibility degrees corresponding to the matrix of nuanced preferences (shown up above). This means, for example, that for player 1 $\varphi\left[R_1\left(s_1^3;s_2^5\right)\right]=LLL$, that is, player 1 sees the strategy (moderate-cooperation, no-cooperation) as having a low degree of feasibility. From these we can evaluate the degree of feasibility of say rule $\Re_{35}$ for player 1. Let us go on to solving the complete LFL-PD game.

3 Linguistic Fuzzy-Logic 2X2 PD Game

Let us consider three different situations depending on the type of player, that is, whether the player is mean-spirited or nice-spirited. Each of these two types is in fact a set of linguistic subtypes assuming a linguistic value – there is graduality in meanness and niceness of the actors. I present the solutions to the game in terms of six sets of couplets. ν_1^k and φ_1^k stand for the nuance and feasibility degrees of an initial choice k for player 1. ν_2^l and φ_2^l stand for the nuance and feasibility degrees of an initial choice l for player 2. $\zeta\left(\nu_1^k\right)$ and $\zeta\left(\varphi_1^k\right)$ stand for the nuance and feasibility degrees of a strategic arrangement k for player 1. $\zeta\left(\nu_2^k\right)$ and $\zeta\left(\varphi_2^k\right)$ stand for the nuance and feasibility degrees of a strategic arrangement l for player 2. The nuance and feasibility degrees of Nash solutions are denoted by $\nu*_{k_1^*l_2^*}^{pr}$ and $\varphi*_{k_1^*l_2^*}^{pr}$. I present the different solutions obtained for various values of orness ω for each actor.

The various values for ω (orness) stand for different ways of combining the feasibility degrees of the possible rules to form a strategic arrangement. ω is a measure of how disjunctively-like (OR-like) or conjunctively-like (AND-like) the various feasibility degrees are combined. As explained in Chapter 2, LOWA is an "orand" operator located somewhere between the "AND" and "OR" logical operations – LOWA has simultaneously a finite degree of "orness" and a finite degree of "andness." Moreover, an or-like LOWA operator with $\omega > 0.5$ describes the case of a player who prefers nuances (i.e., levels) of cooperation with high levels of feasibility. This means that this player will be more trusting and less cautious when deciding on the strategic arrangement. Conversely, an and-like LOWA operator with $\omega > 0.5$ describes the case of a player who prefers nuances of cooperation with low levels of feasibility. This means that this player will be less trusting and more cautious when deciding on the strategic arrangement.

Although this looks similar to the notion of mixed strategies of conventional game theory, it is fundamentally different from it. In a mixed strategy a player would have a probability for using any one of the possible rules, with the understanding that the player might only use ONE such rule at any point in time. In the LFL approach the player will combine more or less the feasibility degrees of all rules to design a strategy. The "more or less" phrase is not to be understood probabilistically, but rather constitutively; that is: as a combination of "bits" from all strategies, a sort of concatenation. The LOWA operator allows us to incorporate vagueness (or fuzzy overlap) in the degree of feasibility of the strategic arrangement. Moreover, the type of Nash equilibrium and its degrees of nuance and feasibility depend on the value of orness. To facilitate a comparison with the crisp PD game, I make the C strategy correspond to a nuance value ν= FFF (and feasibility degree φ= FFF) and D correspond to a nuance value ν= NNN (and feasibility degree φ= FFF). The Nash solution of the crisp PD game is thus DD with a nuance value ν=NNN and a feasibility value φ=FFF. A fully mean-spirited (nice-spirited) player is one who has an FFF (FFF) degree of feasibility for NNN- (FFF-) cooperation and an NNN (NNN) degree of feasibility for FFF- (NNN-) cooperation. That is: a mean-spirited (nice-spirited) player is more prone to see null (full) cooperation as fully (fully) feasible and by the same token is more prone to see full (null) cooperation with a null (null) degree of feasibility. Thus, we have the nuance degrees for a mean-spirited player as ν-cooperation = {FFF, VHH, HHH, MMM, LLL, VLL, NNN} with the corresponding feasibility degrees φ-feasibility = {NNN, VLL, LLL, MMM, HHH, VHH, FFF}. For a nice-spirited player we have: ν-cooperation = {FFF, VHH, HHH, MMM, LLL, VLL, NNN} and the corresponding feasibility degrees φ-feasibility = {FFF, VHH, HHH, MMM, LLL, VLL, NNN}.[3]

We are now in a position to consider in full the 2x2 LFL-PD game. I consider three different situations depending on the type of player, that is, whether the player is mean-spirited or nice-spirited. Each of these two types is a set of linguistic types assuming a linguistic value. In other words, *there is graduality in meanness and niceness of the actors.* I present the solutions to the game in terms of six sets of couplets. ν_1^k and φ_1^k stand for the nuance and feasibility degrees of an initial choice k for player 1, ν_2^l and φ_2^l stand for the nuance and feasibility degrees of

[3] Note that we can have other types of players defined with other possible combinations of nuances and feasibilities, which would not fit in the categories of actors that I use as illustrative examples.

an initial choice l for player 2, $\zeta(\nu_1^k)$ and $\zeta(\varphi_1^k)$ stand for the nuance and feasibility degrees of a strategic arrangement k for player 1, $\zeta(\nu_2^l)$ and $\zeta(\varphi_2^l)$ stand for the nuance and feasibility degrees of a strategic arrangement l for player 2. The nuance and feasibility degrees of Nash solutions are denoted by $\nu *^{pr}_{k_1^* l_2^*}$ and $\varphi *^{pr}_{k_1^* l_2^*}$. I present the different solutions obtained for various values of orness **ω** for each actor. It is clear that the type of Nash equilibrium and its degrees of nuance and feasibility depend on the value of orness.[4] For the sake of comparing with the crisp PD game, the C strategy corresponds to a nuance value ν= FFF (and feasibility degree φ= FFF) and D corresponds to a nuance value ν= NNN (and feasibility degree φ= FFF). The Nash solution of the crisp PD game is thus DD with a nuance value ν=NNN (i.e., null cooperation) and a feasibility value φ=FFF (i.e., fully feasible).

Two mean-spirited players:

I define a mean-spirited player as one who has an FFF degree of feasibility for NNN-cooperation and an NNN degree of feasibility for FFF-cooperation. That is: a mean-spirited player is more prone to see null cooperation (defection) as fully feasible and by the same token is more prone to see full cooperation with a null degree of feasibility, i.e., unfeasible. We thus have the following correspondence in a game with two mean-spirited players (Table 1):

Table 1 Values of nuance and feasibility for PD game with two mean spirited players

1: Mean	ν-cooperation	FFF	VHH	HHH	MMM	LLL	VLL	NNN
	φ-feasibility	NNN	VLL	LLL	MMM	HHH	VHH	FFF
2: Mean	ν-cooperation	FFF	VHH	HHH	MMM	LLL	VLL	NNN
	φ-feasibility	NNN	VLL	LLL	MMM	HHH	VHH	FFF

In this game with **ω=0.1** we have two more or less mean-spirited players, with varying degrees of meanness (Table 2). The level of meanness is defined in terms of how the players perceive cooperation and its feasibility. We can see from Table 2 that there is a large number (14) of F-N Nash equilibria for a small value of orness. These are situations where the players when evaluating the feasibility of

[4] Note that a medium value of $\omega \sim 0.5$ means that the linguistic values being combined are all equally important, whereas for $\omega > 0.5$ ($\omega < 0.5$) the term with the highest (lowest) linguistic value in the set of terms to be combined strongly dominates the combination. LOWA operator with many of the weights near the top will be an ***orlike*** operator with $\omega > 0.5$, while those operators with most of the weights at the bottom will be ***andlike*** operators with $\omega < 0.5$.

Table 2 Highest Nash equilibria for ω=0.1 for PD game with two mean spirited players.

ω=0.1											
Row	v_1^k	φ_1^k	v_2^l	φ_2^l	$\zeta(v_1^k)$	$\zeta(\varphi_1^k)$	$\zeta(v_2^l)$	$\zeta(\varphi_2^l)$	$v^{*pr}_{k_1^* l_2^*}$	$\varphi^{*pr}_{k_1^* l_2^*}$	Nash
1	FFF	NNN	FFF	NNN	FFF	MMM	FFF	MMM	MMM	MMM	F-N
2	FFF	NNN	VHH	VLL	FFF	MMM	VHH	MMM	MMM	MMM	F-N
3	FFF	NNN	NNN	FFF	FFF	MMM	FFF	MMM	MMM	MMM	F-N
4	VHH	VLL	FFF	NNN	VHH	MMM	FFF	MMM	MMM	MMM	F-N
5	VHH	VLL	VHH	VLL	VHH	MMM	VHH	MMM	MMM	MMM	F-N
6	VHH	VLL	MMM	MMM	VHH	MMM	MMM	LLL	MMM	MMM	F-N
7	VHH	VLL	NNN	FFF	VHH	MMM	FFF	MMM	MMM	MMM	F-N
8	MMM	MMM	FFF	NNN	MMM	LLL	FFF	MMM	MMM	MMM	F-N
9	MMM	MMM	VHH	VLL	MMM	LLL	VHH	MMM	MMM	MMM	F-N
10	MMM	MMM	NNN	FFF	MMM	LLL	FFF	MMM	MMM	MMM	F-N
11	NNN	FFF	FFF	NNN	FFF	MMM	FFF	MMM	MMM	MMM	F-N
12	NNN	FFF	VHH	VLL	FFF	MMM	VHH	MMM	MMM	MMM	F-N
13	NNN	FFF	MMM	MMM	FFF	MMM	MMM	LLL	MMM	MMM	F-N
14	NNN	FFF	NNN	FFF	FFF	MMM	FFF	MMM	MMM	MMM	F-N

strategic arrangements of the game favor strongly those with low degrees of feasibility (MMM=moderate or LLL=low as shown in the table). This is displayed in columns 7 and 9 which show the degrees of feasibility of the two players for strategic arrangements that result from initial choices of the players shown in columns 2 and 3 for player 1 and 4 and 5 for player 2. All equilibria have the same degree of nuance and feasibility, i.e., MMM. In other words, the F-N Nash equilibrium is characterized by moderate-cooperation and is perceived to be moderately feasible. This equilibrium can be reached from a variety of combinations of initial choices of the two players. A remarkable result is shown in the last row of the table where F-N Nash equilibrium is obtained from an initial situation where both players perceive non-cooperation (NNN) as fully feasible (FFF). The converse of this situation is shown in row 1 where the two players perceive full-cooperation as not feasible. The situation is repeated in rows 3 and 11. In sum, even when both players initially prefer full defection, playing the LFL game leads to a Nash equilibrium of the strongest type for moderate cooperation! The dilemma of the crisp Boolean logic PD game is lifted in the sense that moderate cooperation is rationally preferred to full defection.

For **ω=0.4**, we obtain an F-N Nash equilibrium for high cooperation with a high degree of feasibility (Table 3). This equilibrium emerges from an initial choice situation where both players perceive full cooperation as not feasible. However, at the level of strategic arrangements involving both players, we find

Table 3 Highest Nash equilibria for ω=0.4 for PD game with two mean spirited players.

ω=0.4											
Row	v_1^k	φ_1^k	v_2^l	φ_2^l	$\zeta(v_1^k)$	$\zeta(\varphi_1^k)$	$\zeta(v_2^l)$	$\zeta(\varphi_2^l)$	$v^{*pr}_{k_1^* l_2^*}$	$\varphi^{*pr}_{k_1^* l_2^*}$	Nash
1	FFF	NNN	FFF	NNN	FFF	HHH	FFF	HHH	HHH	HHH	F-N

Table 4 Highest Nash equilibria for ω=0.8 for PD game with two mean spirited players.

ω=0.8											
Row	v_1^k	φ_1^k	v_2^l	φ_2^l	$\zeta(v_1^k)$	$\zeta(\varphi_1^k)$	$\zeta(v_2^l)$	$\zeta(\varphi_2^l)$	$v*_{k_1^* l_2^*}^{pr}$	$\varphi*_{k_1^* l_2^*}^{pr}$	**Nash**
1	FFF	NNN	FFF	NNN	FFF	HHH	FFF	HHH	HHH	HHH	F-N
2	FFF	NNN	VHH	VLL	FFF	HHH	VHH	HHH	HHH	HHH	F-N
3	FFF	NNN	NNN	FFF	FFF	HHH	VHH	HHH	HHH	HHH	F-N
4	VHH	VLL	FFF	NNN	VHH	HHH	FFF	HHH	HHH	HHH	F-N
5	VHH	VLL	VHH	VLL	VHH	HHH	VHH	HHH	HHH	HHH	F-N
6	VHH	VLL	NNN	FFF	VHH	HHH	VHH	HHH	HHH	HHH	F-N
7	NNN	FFF	FFF	NNN	VHH	HHH	FFF	HHH	HHH	HHH	F-N
8	NNN	FFF	VHH	VLL	VHH	HHH	VHH	HHH	HHH	HHH	F-N
9	NNN	FFF	NNN	FFF	VHH	HHH	VHH	HHH	HHH	HHH	F-N

both strategic arrangements have full cooperation with a high degree of feasibility. The PD dilemma of the crisp Boolean logic game has been lifted. A value of 0.4 for orness means that the players put almost equal emphasis on the feasibility of the possible choices when aggregating the information to evaluate the feasibility of the various strategic arrangements. For orness=0.5 the different choices would be equally weighted in the aggregation process.

Increasing still the degree of orness to 0.8 (Table 4), that is, the players put much more emphasis on the strategic arrangements with high degrees of feasibility as shown in columns 7 and 9, we obtain 9 Nash equilibria, all of them of the F-N Nash equilibrium type. From Table 4 we can see that high cooperation with a high degree of feasibility can be arrived at starting from different initial choices of the individual players, that is, before strategic interaction is taken into account. Rows 1, 3, 7, and 9 show that although both players might perceive full cooperation as not feasible (or equivalently, no cooperation as fully feasible), strategic interaction under linguistic fuzzy logic leads to high cooperation with a high degree of feasibility. There is no dilemma in the game anymore. In sum, in situation where we have two more or less mean-spirited players, cooperation is feasible, at a high level with a high degree of feasibility.

<u>One nice-spirited player against a mean-spirited player</u>

Table 5 Values of nuance and feasibility for PD game with one mean- and one nice-spirited players.

1: Nice	v-cooperation	FFF	VHH	HHH	MMM	LLL	VLL	NNN
	φ-feasibility	FFF	VHH	HHH	MMM	LLL	VLL	NNN
2: Mean	v-cooperation	FFF	VHH	HHH	MMM	LLL	VLL	NNN
	φ-feasibility	NNN	VLL	LLL	MMM	HHH	VHH	FFF

Table 6 Highest Nash equilibria for ω=0.1 for PD game with one mean- and one nice-spirited players.

ω=0.1											
Row	ν_1^k	φ_1^k	ν_2^l	φ_2^l	$\varsigma(\nu_1^k)$	$\varsigma(\varphi_1^k)$	$\varsigma(\nu_2^l)$	$\varsigma(\varphi_2^l)$	$\nu *_{k_1^* l_2^*}^{pr}$	$\varphi *_{k_1^* l_2^*}^{pr}$	Nash
1	FFF	FFF	FFF	NNN	HHH	MMM	FFF	MMM	MMM	MMM	F-N
2	FFF	FFF	VHH	VLL	HHH	MMM	VHH	MMM	MMM	MMM	F-N
3	FFF	FFF	MMM	MMM	HHH	MMM	MMM	LLL	MMM	MMM	F-N
4	FFF	FFF	NNN	FFF	HHH	MMM	VHH	MMM	MMM	MMM	F-N
5	MMM	MMM	FFF	NNN	MMM	LLL	FFF	MMM	MMM	MMM	F-N
6	MMM	MMM	VHH	VLL	MMM	LLL	VHH	MMM	MMM	MMM	F-N
7	MMM	MMM	NNN	FFF	MMM	LLL	VHH	MMM	MMM	MMM	F-N
8	VLL	VLL	FFF	NNN	VHH	MMM	FFF	MMM	MMM	MMM	F-N
9	VLL	VLL	VHH	VLL	VHH	MMM	VHH	MMM	MMM	MMM	F-N
10	VLL	VLL	MMM	MMM	VHH	MMM	MMM	LLL	MMM	MMM	F-N
11	VLL	VLL	NNN	FFF	VHH	MMM	VHH	MMM	MMM	MMM	F-N
12	NNN	NNN	FFF	NNN	FFF	MMM	FFF	MMM	MMM	MMM	F-N
13	NNN	NNN	VHH	VLL	FFF	MMM	VHH	MMM	MMM	MMM	F-N
14	NNN	NNN	MMM	MMM	FFF	MMM	MMM	LLL	MMM	MMM	F-N
15	NNN	NNN	NNN	FFF	FFF	MMM	VHH	MMM	MMM	MMM	F-N

In this game we have a more or less nice-spirited player against a more or less mean-spirited player (Table 5). These players are the mirror images of one another. The level of niceness and meanness are defined in terms of how these players perceive cooperation and its feasibility as shown in Table 5. We can see from Table 6 that there is a large number (15) of F-N Nash equilibria for a small value of orness **ω=0.1**, that is, in situations where the players when evaluating the feasibility of strategic arrangements of the game favor strongly those with low degrees of feasibility (MMM=moderate or LLL=low as shown Table 6). This is displayed in columns 7 and 9 which show the degrees of feasibility of the two players for strategic arrangements that result from initial choices of the players shown in columns 2 and 3 for player 1 and 4 and 5 for player 2. All equilibria have the same degree of nuance and feasibility, i.e., MMM. In other words, the F-N Nash equilibrium is characterized by moderate-cooperation and is perceived to be moderately feasible. This equilibrium can be reached from a variety of combinations of initial choices of the two players. A remarkable result is shown in the last row of the table where F-N Nash equilibrium is obtained from an initial situation where player 2 perceives non-cooperation (NNN) as fully feasible (FFF) whereas player 1 perceives non-cooperation as not feasible. The converse of this situation is shown in the first row where player 1 perceives full-cooperation as fully feasible whereas player 2 perceives full cooperation as not feasible. This situation is repeated in row 4 where player 1 perceives non-cooperation as fully feasible, and in

Table 7 Highest Nash equilibria for ω=0.4 for PD game with one mean- and one nice-spirited players.

ω=0.4											
Row	ν_1^k	φ_1^k	ν_2^l	φ_2^l	$\zeta(\nu_1^k)$	$\zeta(\varphi_1^k)$	$\zeta(\nu_2^l)$	$\zeta(\varphi_2^l)$	$\nu*^{pr}_{k_1^* l_2^*}$	$\varphi*^{pr}_{k_1^* l_2^*}$	Nash
1	VLL	VLL	FFF	NNN	VHH	MMM	FFF	HHH	MMM	HHH	F
2	NNN	NNN	FFF	NNN	FFF	HHH	FFF	HHH	HHH	HHH	F-N
3	NNN	NNN	VHH	VLL	FFF	HHH	VHH	MMM	MMM	HHH	F

row 12 where player 1 perceives non-cooperation as not feasible whereas player 2 perceives full-cooperation as not feasible. In sum, even when one of the players fully defects we still get a Nash equilibrium of the strongest type for moderate cooperation! The dilemma of the crisp Boolean logic PD game is lifted in the sense that moderate cooperation is rationally preferred to full defection.

When the degree of orness is increased from 0.1 to 0.4 we still have moderate cooperation with a high degree of feasibility (Table 7). An F-N Nash equilibrium emerges with a high degree of feasibility. We thus have two F-type Nash equilibrium with moderate cooperation and an F-N Nash equilibrium all with a high degree of feasibility.

Increasing still further the degree of orness to 0.9 (Table 8), that is, the players put much more emphasis on the strategic arrangements with high degrees of feasibility as shown in columns 7 and 9, we obtain 13 Nash equilibria, but only one of

Table 8 Highest Nash equilibria for ω=0.9 for PD game with one mean- and one nice-spirited players.

ω=0.9											
Row	ν_1^k	φ_1^k	ν_2^l	φ_2^l	$\zeta(\nu_1^k)$	$\zeta(\varphi_1^k)$	$\zeta(\nu_2^l)$	$\zeta(\varphi_2^l)$	$\nu*^{pr}_{k_1^* l_2^*}$	$\varphi*^{pr}_{k_1^* l_2^*}$	Nash
1	NNN	NNN	FFF	NNN	FFF	FFF	FFF	FFF	FFF	FFF	F-N
2	FFF	FFF	FFF	NNN	HHH	VHH	FFF	FFF	VHH	FFF	F
3	VHH	VHH	FFF	NNN	HHH	VHH	FFF	FFF	VHH	FFF	F
4	HHH	HHH	FFF	NNN	HHH	HHH	FFF	FFF	HHH	FFF	F
5	MMM	MMM	FFF	NNN	MMM	MMM	FFF	FFF	MMM	FFF	F
6	LLL	LLL	FFF	NNN	HHH	HHH	FFF	FFF	HHH	FFF	F
7	VLL	VLL	FFF	NNN	VHH	VHH	FFF	FFF	VHH	FFF	F
8	NNN	NNN	VHH	VLL	FFF	FFF	VHH	VHH	VHH	FFF	F
9	NNN	NNN	HHH	LLL	FFF	FFF	HHH	HHH	HHH	FFF	F
10	NNN	NNN	MMM	MMM	FFF	FFF	MMM	MMM	MMM	FFF	F
11	NNN	NNN	LLL	HHH	FFF	FFF	HHH	HHH	HHH	FFF	F
12	NNN	NNN	VLL	VHH	FFF	FFF	VHH	VHH	VHH	FFF	F
13	NNN	NNN	NNN	FFF	FFF	FFF	VHH	VHH	VHH	FFF	F

them is a F-N Nash equilibrium. The latter is the result of a scenario where player 1 perceives non-cooperation as not feasible and player 2 perceives full cooperation as not feasible. This however does not ineluctably lead to full defection as the crisp Boolean logic PD game would predict. From the table we can see that moderate, high, very high and full cooperation are all fully feasible. Thus the players can in the worst case scenario have moderate cooperation fully feasible. There is no dilemma in the game anymore. In sum, in situation where we have one nice-spirited player against a mean-spirited one, cooperation is feasible, at least at a moderate level with a moderate degree of feasibility.

Two nice-spirited players

Table 9 Values of nuance and feasibility for PD game with two nice-spirited players.

1: Nice	ν-cooperation	FFF	VHH	HHH	MMM	LLL	VLL	NNN
	φ-feasibility	FFF	VHH	HHH	MMM	LLL	VLL	NNN
2: Nice	ν-cooperation	FFF	VHH	HHH	MMM	LLL	VLL	NNN
	φ-feasibility	FFF	VHH	HHH	MMM	LLL	VLL	NNN

Much of what was said in the two previous cases remains valid for the case where two more or less nice-spirited players in the game (Table 9).

Table 10 Highest Nash equilibria for ω=0.1 for PD game with two nice-spirited players.

ω=0.1											
Row	ν_1^k	φ_1^k	ν_2^l	φ_2^l	$\zeta(\nu_1^k)$	$\zeta(\varphi_1^k)$	$\zeta(\nu_2^l)$	$\zeta(\varphi_2^l)$	$\nu^{*pr}_{k_1^* l_2^*}$	$\varphi^{*pr}_{k_1^* l_2^*}$	**Nash**
1	FFF	FFF	FFF	FFF	HHH	MMM	HHH	MMM	MMM	MMM	F-N
2	FFF	FFF	MMM	MMM	HHH	MMM	MMM	LLL	MMM	MMM	F-N
3	FFF	FFF	VLL	VLL	HHH	MMM	VHH	MMM	MMM	MMM	F-N
4	FFF	FFF	NNN	NNN	HHH	MMM	FFF	MMM	MMM	MMM	F-N
5	MMM	MMM	FFF	FFF	MMM	LLL	HHH	MMM	MMM	MMM	F-N
6	MMM	MMM	VLL	VLL	MMM	LLL	VHH	MMM	MMM	MMM	F-N
7	MMM	MMM	NNN	NNN	MMM	LLL	FFF	MMM	MMM	MMM	F-N
8	VLL	VLL	FFF	FFF	VHH	MMM	HHH	MMM	MMM	MMM	F-N
9	VLL	VLL	MMM	MMM	VHH	MMM	MMM	LLL	MMM	MMM	F-N
10	VLL	VLL	VLL	VLL	VHH	MMM	VHH	MMM	MMM	MMM	F-N
11	VLL	VLL	NNN	NNN	VHH	MMM	FFF	MMM	MMM	MMM	F-N
12	NNN	NNN	FFF	FFF	FFF	MMM	HHH	MMM	MMM	MMM	F-N
13	NNN	NNN	MMM	MMM	FFF	MMM	MMM	LLL	MMM	MMM	F-N
14	NNN	NNN	VLL	VLL	FFF	MMM	VHH	MMM	MMM	MMM	F-N
15	NNN	NNN	NNN	NNN	FFF	MMM	FFF	MMM	MMM	MMM	F-N

When the level of orness = 0.1 (Table 10), we have 15 F-N Nash equilibria, which are arrived at in different ways, that is, starting with various individual choices before the onset of strategic interaction. Just as in the two previous situations, we have an F-N Nash equilibrium with a moderate degree of cooperation and a moderate level of feasibility.

Table 11 Highest Nash equilibria for ω=0.4 for PD game with two nice-spirited players.

ω=0.4											
Row	v_1^k	φ_1^k	v_2^l	φ_2^l	$\zeta(v_1^k)$	$\zeta(\varphi_1^k)$	$\zeta(v_2^l)$	$\zeta(\varphi_2^l)$	$v*^{pr}_{k_1^* l_2^*}$	$\varphi*^{pr}_{k_1^* l_2^*}$	Nash
1	VLL	VLL	NNN	NNN	HHH	MMM	HHH	HHH	MMM	HHH	F
2	NNN	NNN	VLL	VLL	HHH	HHH	HHH	MMM	MMM	HHH	F
3	NNN	NNN	NNN	NNN	HHH	HHH	HHH	HHH	HHH	HHH	F-N

For intermediate degrees of orness (Table 11), we also obtain three Nash equilibria, two of F type and one of F-N type. While all three equilibria have a high degree of feasibility, the F-N one stands for high cooperation whereas the two others stand for moderate cooperation. There is no dilemma in the game.

For orness=0.9 (Table 12), we obtain 13 Nash equilibria, all fully feasible, but only the F-N one stands for full cooperation, whereas we find 6 equilibria with a very high level of cooperation, 4 equilibria with a high level of cooperation, and 2

Table 12 Highest Nash equilibria for ω=0.9 for PD game with two nice-spirited players.

ω=0.9											
Row	v_1^k	φ_1^k	v_2^l	φ_2^l	$\zeta(v_1^k)$	$\zeta(\varphi_1^k)$	$\zeta(v_2^l)$	$\zeta(\varphi_2^l)$	$v*^{pr}_{k_1^* l_2^*}$	$\varphi*^{pr}_{k_1^* l_2^*}$	Nash
1	FFF	FFF	NNN	NNN	HHH	VHH	FFF	FFF	VHH	FFF	F
2	VHH	VHH	NNN	NNN	HHH	VHH	FFF	FFF	VHH	FFF	F
3	HHH	HHH	NNN	NNN	HHH	HHH	FFF	FFF	HHH	FFF	F
4	MMM	MMM	NNN	NNN	MMM	MMM	FFF	FFF	MMM	FFF	F
5	LLL	LLL	NNN	NNN	HHH	HHH	FFF	FFF	HHH	FFF	F
6	VLL	VLL	NNN	NNN	VHH	VHH	FFF	FFF	VHH	FFF	F
7	NNN	NNN	FFF	FFF	FFF	FFF	HHH	VHH	VHH	FFF	F
8	NNN	NNN	VHH	VHH	FFF	FFF	HHH	VHH	VHH	FFF	F
9	NNN	NNN	HHH	HHH	FFF	FFF	HHH	HHH	HHH	FFF	F
10	NNN	NNN	MMM	MMM	FFF	FFF	MMM	MMM	MMM	FFF	F
11	NNN	NNN	LLL	LLL	FFF	FFF	HHH	HHH	HHH	FFF	F
12	NNN	NNN	VLL	VLL	FFF	FFF	VHH	VHH	VHH	FFF	F
13	NNN	NNN	NNN	NNN	FFF	FFF	FFF	FFF	FFF	FFF	F-N

equilibria with a moderate level of cooperation. Hence, the lowest level of cooperation which is fully feasible is moderate. There is no dilemma in the game.

To sum up: in all three situations – two mean-spirited players, two nice-spirited players, one mean-spirited player against one nice-spirited player – there is no dilemma in the LFL-PD game. In addition, there at least is one strong F-N Nash equilibrium, and the lowest level of cooperation is moderate with a moderate level of feasibility. In some cases, we have full cooperation with a full degree of feasibility. In the next section, I illustrate the power of using a LFL-game in shedding light on the role of reassurance in international relations.

4 Role of Reassurance in International Relations

Andrew Kydd (2000) used game theory to develop a costly-signaling theory of reassurance, which focuses on the sending and interpretation of costly signals, as a way of resolving the problem of mistrust in IR. Signals are made costly enough to persuade the receiver that the sender is trustworthy (2000:326). In this pursuit, Kydd modified the usual trust game modeled using a PD game by dividing the game into two rounds, a lesser initial round, followed by a final, more important round. The actors can hence predicate their choices in the second round on what happened in the first one. Kydd introduces a parameter $0 < \alpha < 1$ as a measure of signal costliness. Round 1 of the game is worth α while round 2 is worth $1\text{-}\alpha$. The most promising equilibrium – termed as separating – occurs over an interval of α^* values. Kydd shows that in this parameter space cooperation would occur at much lower levels of trust than in a one-round trust game. "The signals must be costly, but not too costly. Make them too easy and they become cheap talk ... Such claims are unpersuasive because there is nothing to prevent untrustworthy types from making them. However, the signals cannot be made too costly either, or the nice types will be too fearful to send them … cooperation may be possible for levels of trust that are quite low – low enough to preclude cooperation in the simple one-round trust game" (2000:340). As an illustration of his game-theoretic argument Kydd looks at the negotiations that led to the end of the Cold War, arguing that Gorbachev engaged the US leaders into a game of reassurance. The latter thus succeeded in fostering trust when the signals (i.e., unilateral concessions made by Gorbachev) became truly costly for the Soviet Union (2000:341).

I address the same issue but using an approach based on LFL-game theory. Following Kydd's approach, I allow the players to play two consecutive PD games, that is, the second game takes as initial choices of the players the outcome of the first game. One key difference with Kydd's game is that I do not use signaling costs as a way of conveying the actors' levels of readiness to cooperate with one another. I instead use the notion of feasibility in combination with the notion of nuanced cooperation (or defection) to describe how the actors communicate with one another. For example, Kydd (2000:338) argues that "The secret to making the separating equilibrium work is finding a signal that is adequately costly to deter the mean types from sending it but not so costly that the nice type is afraid to send it." The LFL-approach counterpart would be: *the secret to making reassurance work is that the actors choose strategies that have higher degrees of feasibility although the corresponding degrees of nuanced cooperation might not be the highest one available. Conversely, the equivalent of a low cost signal will be to choose a strategy that has a higher level of nuanced cooperation but with a low level of feasibility.* This differs from Kydd's Boolean logic approach on two main points. First, the actors are not just looking for cooperation or defection. They are instead looking for more or less cooperation. They want to know: how nuanced will their mutual cooperation be? Is it very weak or rather strong or somewhere in between? Second, how feasible are such levels of cooperation? Undoubtedly, these degrees of feasibility are shaped by cost analysis and cost signaling. However, the notion of feasibility is not restricted to these conventional measures for it includes the actors' perceptions of the past, present, and the future not only in terms of costs but also in terms of intangibles such as identity (and emotions). As explained in the previous section, in the LFL game actors have a propensity toward seeking more or less feasible strategies, an aspect of the game which is modeled by the level of orness ω. The actors thus inquire: At what levels of orness would an actor have a propensity of favoring low (high) degrees of feasibility? In the two-round version of the reassurance game the actors would then have two different levels of orness ω_1 and ω_2, one for each round.

Following Kydd's model one would assume that $\omega_1 + \omega_2 = 1$. That is: the propensity toward choosing more or less feasible strategies in the two rounds of the game are mutually compensatory – a propensity to favor high levels of feasibility during round one would be compensated with a propensity to favor low levels of feasibility during round two, and vice versa. In this interpretation the degree of feasibility becomes the equivalent of cost in Kydd's formulation. However, this is only one possible combination of the degrees of orness of the two rounds. Indeed, in the LFL game we have two degrees of freedom: the level or nuance of strategies and the degree of feasibility of these nuanced strategies. We do not just have cooperation and defection which are either chosen or rejected due to their cost/benefit. This opens up the possibility of many more ways of combining the propensities (i.e., levels of orness) of the two rounds. For example, at the end of round one, depending on the value of orness ω_1, a player possesses a number of choices all of which are equilibria (either of N, F, or F-N types as explained in the previous section). These equilibria produce at least moderate (MMM) cooperation which is at least highly (HHH) feasible. As shown in Table 13, I find at the end of round 1 moderate cooperation with a high degree of feasibility for $\omega_1 \leq 0.2$, high cooperation with a high degree of feasibility for $0.2 \leq \omega_1 \leq 0.8$, and full cooperation with a full degree of feasibility for $0.8 \leq \omega_1$.

Translated in policy terms, these results mean that negotiators do not have just to cooperate or defect; they can engage one another at various levels of cooperation which they believe as more or less feasible! Therefore, the negotiators can move on to the second round of the reassurance game along many different equilibrium paths, each characterized with a level of cooperation and a corresponding degree of feasibility. The degrees of nuance and feasibility for the game profile at the end of the second round are shown in Table 13, where ω_1 and ω_2 stand for orness levels for rounds one and two. One notable result is that very high (ν_{GP}=VHH) cooperation is fully (φ_{GP}=FFF) feasible at the end of the second round at any orness level of the first round (note that we get FFF-feasible, FFF-cooperation for ω_2 >0.90). In other words: Even starting with (moderately or highly feasible) moderate levels of cooperation at the end of round one can produce fully feasible, very high or full cooperation at the end of round two.

In terms of practical application, this opens up the possibility for continuing the negotiations even when the players are not still fully committed to full cooperation, or even when they do not see full cooperation as fully feasible at the end of the first round of negotiations. We can also interpret this in terms of path-dependency reasoning; that is: moderately feasible, moderate cooperation at the end of the first round puts the players on a path-dependent equilibrium-path which self-feeds to preserve the momentum of cooperation going on. This can also be phrased in terms of "cheap talk" jargon used in conventional game theory. LFL game theory shows that even "cheap talks" can build enough momentum for full cooperation in later stages. This runs into contradiction with Kydd's (2000: 343) argument that, when Gorbachev began to make some arms control advances toward the US such as suggesting a moratorium on nuclear testing and one on SS20 intermediate-range ballistic missile (IRBM) deployments, "these gestures can be usefully conceived of as signals that fall short of the crucial level … these signals were 'cheap talk' rather than 'costly signals.' These moves were ineffective in changing opinions in the West." In light of LFL game theory, I disagree with Kydd's assessment of this historical event and instead concur with Collins (1998:205, cited by Kydd) and Joshua Goldstein and John Freeman (1990:116, cited by Kydd) who reason that although there was no direct response from the US, Gorbachev's gestures did initiate a path-dependent process in a more subtle way, thereby setting the stage for upcoming important gestures and explicit cooperation. Instead of arguing as Kydd did that "these signals failed, largely because they were regarded as moves that did not really hurt the Soviets; that is, they were signals with little or no cost," I think that these events showed that cooperation, if only of a moderate level, can be more or less feasible.

In this respect, we can see how a fuzzy logic approach leads to another important difference with Boolean-logic game theory. For instance, Kydd's conceptualization of the notion of threshold at which point signals become costly is a "yes/no" one. That is: signals remain "cheap talks" until they reach the threshold after which they become effectively credible. Thresholds in fuzzy logic are (of course!) fuzzy – they are continuous fuzzy boundaries between nuanced levels of signaling and cooperation instead of being step-like changes. Even in the cursory analysis that Kydd (understandably) offers from the rise of Gorbachev to the final agreements we can discern a continuous, fuzzy momentum of negotiations from one round to the next, with every new round producing higher levels of cooperation than previous ones. This fits very well with the above briefly discussed LFL game of reassurance. In fact, taking both sides of the negotiations together it is very hard to see (except through *post hoc* reassessment) when was the precise threshold that Kydd argues had occurred. At any point in time the actors not only kept coming back to the table of negotiations but also kept cooperating even more. More importantly, they truly believed increasingly more that *higher levels of co-operation* were *more feasible*. One can even venture to say that the Soviet "New

Table 13 Profile of Reassurance Game

ROUND ONE OF THE GAME

ω_1	0.1	0.2	0.3	0.4	0.5	0.6	0.7	0.8	0.9
Degrees of Nuance: ν_{GP}	MMM	HHH	HHH	HHH	HHH	HHH	HHH	HHH	FFF
Degrees of Feasibility: φ_{GP}	HHH	HHH	HHH	HHH	HHH	HHH	HHH	HHH	FFF

ROUND TWO OF THE GAME

Degrees of Nuance: ν_{GP}

ω_1 \ ω_2	0.1	0.2	0.3	0.4	0.5	0.6	0.7	0.8	0.9
0.1	MMM	MMM	MMM	MMM	MMM	HHH	HHH	HHH	VHH
0.2	MMM	MMM	MMM	MMM	MMM	HHH	HHH	HHH	VHH
0.3	MMM	MMM	HHH	HHH	HHH	HHH	HHH	HHH	VHH
0.4	MMM	MMM	MMM	MMM	MMM	HHH	HHH	HHH	VHH
0.5	MMM	MMM	MMM	MMM	MMM	HHH	HHH	HHH	VHH
0.6	MMM	MMM	MMM	MMM	MMM	HHH	HHH	HHH	VHH
0.7	MMM	MMM	MMM	MMM	MMM	MMM	HHH	HHH	VHH
0.8	MMM	MMM	MMM	MMM	MMM	MMM	HHH	HHH	VHH
0.9	MMM	MMM	MMM	MMM	MMM	MMM	MMM	HHH	VHH

Degrees of Feasibility: φ_{GP}

ω_1 \ ω_2	0.1	0.2	0.3	0.4	0.5	0.6	0.7	0.8	0.9
0.1	MMM	MMM	MMM	MMM	MMM	HHH	HHH	HHH	FFF
0.2	MMM	MMM	MMM	MMM	MMM	HHH	HHH	HHH	FFF
0.3	HHH	HHH	HHH	HHH	HHH	HHH	HHH	HHH	FFF
0.4	HHH	HHH	HHH	HHH	HHH	HHH	HHH	HHH	FFF
0.5	HHH	HHH	HHH	HHH	HHH	HHH	HHH	HHH	FFF
0.6	MMM	MMM	MMM	MMM	MMM	HHH	HHH	HHH	FFF
0.7	MMM	MMM	MMM	MMM	MMM	HHH	HHH	HHH	FFF
0.8	MMM	MMM	MMM	MMM	MMM	MMM	HHH	HHH	FFF
0.9	MMM	MMM	MMM	MMM	MMM	MMM	HHH	HHH	FFF

Thinking" was precisely about believing and being able to convince the West to believe that consecutively higher levels of cooperation between the East and the West were possible with increasingly higher degrees of feasibility. I hence somewhat agree with Kydd's (2000:350) rendering of the end of the Cold War when he says that "one can observe a series of costly signals leading to mutual trust between former adversaries. The attitudes of Western leaders, press, and publics toward the Soviet Union all underwent a substantial transformation." However, his Boolean-logic game theoretic analysis stops short in seeing that there was a continuous progression in the degrees of feasibility of, and levels of, cooperation that underpinned the evolution of the New Thinking and consequent transformation of Western attitudes and strategies from very low levels of feasible cooperation to fully feasible, full cooperation. To sum up this very brief fuzzy-logic rendering of the end of the Cold War, Kydd's analysis is not wrong but misses much of the *fuzzy* dynamics that underpinned the progression from (fully feasible) no-cooperation to fully feasible, full cooperation. I understand that much more empirical work is needed to really make this case than what this discussion allows for. Yet, I believe that LFLA has much to offer in terms of explaining the fuzzy-logical nature of major historical events of our times. In the next section, I continue exploring the problem of trust based on a different formulation of the strategic situation.

5 Linguistic Fuzzy-Logic Game of Trust

Most game-theoretic studies of trust use an extended form of the game. In this chapter I instead use the normal form of a game. In this pursuit, I combine two different standard games, these are: Stag Hunt and Prisoner's Dilemma games.[5] That is, the two actors play two different games. The potentially trusting player SH plays the Stag Hunt game while the would-be trusted player PD plays the Prisoner's Dilemma game. The SH player is faced with two choices T (trust) or NT (not trust) and does not have a dominant strategy. The PD player is faced with two choices NE (not exploit) or E (exploit). PD does have a dominant strategy, E (exploit). Thus, PD would always rationally play E rather than NE due to the gain in payoff when playing E. The logic of the trust game is obtained when the two games are joined together (Figure 2).

Clearly, the outcome simultaneously best for both players is (α,γ), which can only be achieved when SH plays T (trust) and PD plays NE (not exploit). However, for PD it is rational to opt for the dominant strategy E (exploit) which would force SH to opt for NT (not trust). This produces the outcome $(\beta,\beta)<(\alpha,\gamma)$.

---

[5] In a generic symmetric Stag Hunt game, the choices are either to hunt a stag or to hunt a hare. It takes both players cooperating to effectively hunt a stag, while one player acting independently can hunt a hare. Catching a stag brings greater payoff. There are two equilibria in this game: both players hunt a stag and both hunt a hare. The first equilibrium carries the higher payoff and is said to be *payoff dominant*; the second equilibrium carries the least risk and is said to be *risk dominant*.

		PD: NE (not exploit)	PD: E (exploit)
SH	T (trust)	(α, γ)	(δ, α)
	NT (not trust)	(γ, δ)	(β, β)

Fig. 2 Crisp Game of Trust ($\alpha > \gamma > \beta > \delta$)

There is only one pure strategy Nash equilibrium and no Nash equilibrium emerges in the space of mixed strategies when $\alpha > \gamma > \beta > \delta$. The trust dilemma remains unresolved even in the set of mixed strategies. To obtain the LFL-game of trust we generalize the ordinal game which has the following normal form (Figure 3).

		PD: NE (not exploit)	PD: E (exploit)
SH	T (trust)	(1;2)	(4;1)
	NT (not trust)	(2;4)	(3;3)

Fig. 3 Ordinal Crisp Game of Trust

In the case of the crisp game of trust, SH would have two crisp choices, T and NT. From these two crisp choices we would have one set of LFL-choices obtained by applying the elements of a hedge algebra set H to the crisp choice, T. If we suppose for example that the hedge set is given by: H= {NNN, VLL, LLL, MMM, HHH, VHH, FFF}. SH would have the following set of LFL-choices:

$$\{\nu\text{-T} \mid \nu \in H\}=\{ NNN, VLL, LLL, MMM, HHH, VHH, FFF\}\text{-trust} \quad (70)$$

Each of these LFL-choices would have a feasibility degree

$$\varphi \in \{ NNN, VLL, LLL, MMM, HHH, VHH, FFF\}\text{-feasibility} \quad (71)$$

For the PD player, we would have the following set of LFL-choices:

$$\{\nu\text{-E} \mid \nu \in H\}=\{ NNN, VLL, LLL, MMM, HHH, VHH, FFF\}\text{-exploit} \quad (72)$$

Each of these LFL-choices would have a degree of feasibility:

$$\varphi \in \{ NNN, VLL, LLL, MMM, HHH, VHH, FFF\}\text{-feasibility} \quad (73)$$

The preference matrix for a crisp game of trust would read using the following ordinal preferences (with best = 1 and worst = 4):

$$R_{SH}^{Trust} = \begin{array}{c} \\ T \\ NT \end{array} \overset{\begin{array}{cc} NE & E \end{array}}{\begin{bmatrix} 1 & 4 \\ 2 & 3 \end{bmatrix}} \; ; \; R_{PD}^{Trust} = \begin{array}{c} \\ T \\ NT \end{array} \overset{\begin{array}{cc} NE & E \end{array}}{\begin{bmatrix} 2 & 1 \\ 4 & 3 \end{bmatrix}} \tag{74}$$

That is: player SH's preferred ranking is: (T,NE) > (NT,NE) > (NT,E) > (T,E), whereas player PD's preferred ranking is: (T,E) > (T,NE) > (NT,E) > (NT,NE). Because only PD player has a dominant strategy (E), there is no Nash equilibrium in the space of crisp mixed strategies.

To keep the connection with the Boolean-logic ordinal game of trust we need to remember that for the latter we have all degrees of feasibility equal to TRUE (or, in LFL denotation, FFF). The players' rankings can be, for example, translated into the following degrees of nuances in LFL terminology:

- Player SH: FFF for (T,NE), HHH for (NT,NE), LLL for (NT,E), and NNN for (T,E)
- Player PD: FFF for (T,E), HHH for (T,NE), LLL for (NT,E), and NNN for (NT,NE)

We thus have the following two matrices for both players in the crisp ordinal game of trust:

Matrix of Preference Nuances

$$\text{Player SH: } \varphi\left[R_{SH}^{trust}\right] = \begin{bmatrix} FFF & FFF \\ FFF & FFF \end{bmatrix}; \text{ Player PD: } \varphi\left[R_{PD}^{trust}\right] = \begin{bmatrix} FFF & FFF \\ FFF & FFF \end{bmatrix} \tag{75}$$

Matrix of Preference Feasibilities

$$\text{Player SH: } \nu\left[R_{SH}^{trust}\right] = \begin{bmatrix} FFF & NNN \\ HHH & LLL \end{bmatrix}; \text{ Player PD: } \nu\left[R_{PD}^{trust}\right] = \begin{bmatrix} HHH & FFF \\ NNN & LLL \end{bmatrix} \tag{76}$$

In the LFL-PD game we would, for example, write the matrix of nuanced preference:

$$\text{Player SH: } \nu\left[R_{SH}\right] = \begin{bmatrix} FFF & VHH & HHH & MMM & LLL & VLL & NNN \\ FFF & VHH & HHH & MMM & LLL & VLL & NNN \\ VHH & VHH & HHH & MMM & LLL & VLL & VLL \\ VHH & VHH & HHH & MMM & LLL & VLL & VLL \\ VHH & HHH & MMM & MMM & MMM & LLL & LLL \\ HHH & HHH & MMM & MMM & MMM & LLL & LLL \\ HHH & HHH & MMM & MMM & MMM & LLL & LLL \end{bmatrix} \tag{77}$$

Player PD:

$$\nu\left[R_{PD}\right]=\begin{bmatrix} HHH & HHH & VHH & VHH & VHH & FFF & FFF \\ HHH & HHH & HHH & HHH & HHH & VHH & FFF \\ MMM & MMM & MMM & MMM & MMM & HHH & VHH \\ LLL & LLL & LLL & LLL & LLL & MMM & HHH \\ VLL & VLL & VLL & VLL & VLL & LLL & MMM \\ NNN & NNN & VLL & VLL & VLL & LLL & LLL \\ NNN & NNN & VLL & VLL & VLL & LLL & LLL \end{bmatrix} \quad (78)$$

We also need a matrix of feasibility degrees corresponding to the matrix of nuanced preferences. We could, for example, have:

Players SH & PD:

$$\varphi\left[R_{SH}\right]=\varphi\left[R_{PD}\right]=\begin{bmatrix} FFF & VHH & HHH & MMM & HHH & VHH & FFF \\ VHH & HHH & MMM & LLL & MMM & HHH & VHH \\ HHH & MMM & LLL & VLL & LLL & MMM & HHH \\ MMM & LLL & VLL & NNN & VLL & LLL & MMM \\ HHH & MMM & LLL & VLL & LLL & MMM & HHH \\ VHH & HHH & MMM & LLL & MMM & HHH & VHH \\ FFF & VHH & HHH & MMM & HHH & VHH & FFF \end{bmatrix} \quad (79)$$

Let us now consider situations where the degrees of feasibility of the player's initial choices are allowed to vary gradually from null to full (NNN to FFF). We consider four different situations: SH naïve-spirited and PD nice-spirited, SH cautious-spirited and PD nice-spirited, SH naïve-spirited and PD mean-spirited, and SH cautious-spirited and PD mean-spirited. For each game we consider various degrees of orness. In all cases we obtain at least one F-N Nash equilibrium. This means that there is always at least one very stable solution to the LFL-game of trust.

<u>SH cautious-spirited and PD nice-spirited:</u>

Table 14 Values of nuance and feasibility for trust game with SH cautious-spirited and PD nice-spirited players

SH: Cautious	ν-trust	FFF	VHH	HHH	MMM	LLL	VLL	NNN
	φ-feasibility	NNN	VLL	LLL	MMM	HHH	VHH	FFF
PD: Nice	ν-exploit	FFF	VHH	HHH	MMM	LLL	VLL	NNN
	φ-feasibility	NNN	VLL	LLL	MMM	HHH	VHH	FFF

In this situation, SH player is taken to be cautious-spirited, that is, SH is willing to consider full trust as not feasible, while at the same keeping the option open to lower levels of trust with lower degrees of feasibility. SH nonetheless perceives full trust as not feasible. Player PD is taken to be nice-spirited, that is, PD perceives full exploitation as not feasible, while PD perceives lower levels of exploitation feasible with higher degrees of feasibility. PD can get to an extent as to perceiving no exploitation as fully feasible. Table 14 displays the possible levels

Table 15 Highest Nash equilibria for ω=0.1 for trust game with SH cautious-spirited and PD nice-spirited players.

ω=0.1

Row	v_1^k	φ_1^k	v_2^l	φ_2^l	$\zeta(v_1^k)$	$\zeta(\varphi_1^k)$	$\zeta(v_2^l)$	$\zeta(\varphi_2^l)$	$v^{*pr}_{k_1^* l_2^*}$	$\varphi^{*pr}_{k_1^* l_2^*}$	Nash
1	FFF	NNN	FFF	NNN	FFF	MMM	FFF	MMM	MMM	MMM	F-N
2	FFF	NNN	VHH	VLL	FFF	MMM	VHH	MMM	MMM	MMM	F-N
3	FFF	NNN	MMM	MMM	FFF	MMM	MMM	LLL	MMM	MMM	F-N
4	FFF	NNN	LLL	HHH	FFF	MMM	MMM	LLL	MMM	MMM	F-N
5	FFF	NNN	VLL	VHH	FFF	MMM	LLL	LLL	MMM	MMM	F-N
6	FFF	NNN	NNN	FFF	FFF	MMM	LLL	MMM	MMM	MMM	F-N
7	VHH	VLL	FFF	NNN	VHH	MMM	FFF	MMM	MMM	MMM	F-N
8	VHH	VLL	VHH	VLL	VHH	MMM	VHH	MMM	MMM	MMM	F-N
9	VHH	VLL	MMM	MMM	VHH	MMM	MMM	LLL	MMM	MMM	F-N
10	VHH	VLL	LLL	HHH	VHH	MMM	MMM	LLL	MMM	MMM	F-N
11	VHH	VLL	VLL	VHH	VHH	MMM	LLL	LLL	MMM	MMM	F-N
12	VHH	VLL	NNN	FFF	VHH	MMM	LLL	MMM	MMM	MMM	F-N
13	MMM	MMM	FFF	NNN	MMM	LLL	FFF	MMM	MMM	MMM	F-N
14	MMM	MMM	VHH	VLL	MMM	LLL	VHH	MMM	MMM	MMM	F-N
15	MMM	MMM	NNN	FFF	MMM	LLL	LLL	MMM	MMM	MMM	F-N
16	NNN	FFF	FFF	NNN	HHH	MMM	FFF	MMM	MMM	MMM	F-N
17	NNN	FFF	VHH	VLL	HHH	MMM	VHH	MMM	MMM	MMM	F-N
18	NNN	FFF	MMM	MMM	HHH	MMM	MMM	LLL	MMM	MMM	F-N
19	NNN	FFF	LLL	HHH	HHH	MMM	MMM	LLL	MMM	MMM	F-N
20	NNN	FFF	VLL	VHH	HHH	MMM	LLL	LLL	MMM	MMM	F-N
21	NNN	FFF	NNN	FFF	HHH	MMM	LLL	MMM	MMM	MMM	F-N

(nuances) of trust for player SH with corresponding degrees of feasibility and the possible levels of exploitation for player PD with corresponding degrees of feasibility.

For very low degrees of orness **ω=0.1**, I obtain 21 F-N Nash equilibria, all with a moderate level of trust and moderately feasible (Table 15). What is remarkable is that when SH initially perceives full trust as not feasible (such as in rows 1 – 6) we obtain a F-N solution with any of the possible initial choices for player PD. In other words, that player PD is nice-spirited facilitates the emergence of trust and that the logic of the game is LFL makes a certain level of trust possible. The same is true in situations where player SH initially perceives no trust as fully feasible (rows 16 – 21), very high trust with a very low degree of feasibility (rows 7 – 12), or moderate trust as moderately feasible (rows 13 – 15).

For intermediate values of orness **ω=0.6**, we obtain 2 F-N Nash equilibria and five F Nash equilibria (Table 16). All equilibria have a high degree of feasibility.

Table 16 Highest Nash equilibria for ω=0.6 for trust game with SH cautious-spirited and PD nice-spirited players.

ω=0.6

Row	V_1^k	φ_1^k	V_2^l	φ_2^l	$\zeta(V_1^k)$	$\zeta(\varphi_1^k)$	$\zeta(V_2^l)$	$\zeta(\varphi_2^l)$	$V*^{pr}_{k_1^* l_2^*}$	$\varphi*^{pr}_{k_1^* l_2^*}$	Nash
1	FFF	NNN	FFF	NNN	FFF	HHH	FFF	HHH	HHH	HHH	F-N
2	FFF	NNN	NNN	FFF	FFF	HHH	LLL	MMM	HHH	HHH	F-N
3	FFF	NNN	VHH	VLL	FFF	HHH	VHH	MMM	MMM	HHH	F
4	FFF	NNN	HHH	LLL	FFF	HHH	VHH	MMM	MMM	HHH	F
5	VHH	VLL	FFF	NNN	VHH	MMM	FFF	HHH	MMM	HHH	F
6	HHH	LLL	FFF	NNN	VHH	MMM	FFF	HHH	MMM	HHH	F
7	NNN	FFF	FFF	NNN	HHH	MMM	FFF	HHH	MMM	HHH	F

Table 17 Highest Nash equilibria for ω=0.9 for trust game with SH cautious-spirited and PD nice-spirited players.

ω=0.9

Row	V_1^k	φ_1^k	V_2^l	φ_2^l	$\zeta(V_1^k)$	$\zeta(\varphi_1^k)$	$\zeta(V_2^l)$	$\zeta(\varphi_2^l)$	$V*^{pr}_{k_1^* l_2^*}$	$\varphi*^{pr}_{k_1^* l_2^*}$	Nash
1	FFF	NNN	FFF	NNN	FFF	FFF	FFF	FFF	FFF	FFF	F-N
2	FFF	NNN	VHH	VLL	FFF	FFF	VHH	VHH	VHH	FFF	F
3	FFF	NNN	HHH	LLL	FFF	FFF	HHH	HHH	HHH	FFF	F
4	FFF	NNN	MMM	MMM	FFF	FFF	MMM	MMM	MMM	FFF	F
5	FFF	NNN	LLL	HHH	FFF	FFF	MMM	HHH	HHH	FFF	F
6	FFF	NNN	VLL	VHH	FFF	FFF	LLL	VHH	VHH	FFF	F
7	FFF	NNN	NNN	FFF	FFF	FFF	LLL	VHH	VHH	FFF	F
8	VHH	VLL	FFF	NNN	VHH	VHH	FFF	FFF	VHH	FFF	F
9	HHH	LLL	FFF	NNN	HHH	HHH	FFF	FFF	HHH	FFF	F
10	MMM	MMM	FFF	NNN	MMM	MMM	FFF	FFF	MMM	FFF	F
11	LLL	HHH	FFF	NNN	HHH	HHH	FFF	FFF	HHH	FFF	F
12	VLL	VHH	FFF	NNN	HHH	VHH	FFF	FFF	VHH	FFF	F
13	NNN	FFF	FFF	NNN	HHH	VHH	FFF	FFF	VHH	FFF	F

The F-N ones stand for high level of trust, whereas the F ones stand for moderate level of trust.

In situations with high values of orness **ω=0.9**, we obtain 13 Nash equilibria, one of which is of F-N type while the remaining ones are of F type (Table 17). All equilibria are fully feasible. The F-N Nash equilibrium stands for full trust. The F ones are distributed as: 6 with very high degree of trust, 3 with high degree of trust, and two with moderate degree of trust. What is remarkable is that in the F-N case, player SH initially perceives full trust as not feasible. However, playing the game according to LFL rules and assuming that the PD player initially perceives full exploitation as not feasible leads to full trust which is fully feasible.

SH cautious-spirited and PD mean-spirited:

In this situation, SH is cautious-spirited and PD is mean spirited. That is, SH initially perceives full trust as not feasible and no trust as fully feasible, and PD initially perceives full exploitation as fully feasible and no exploitation as not feasible. Both actors can initially perceive other levels of trust (player SH) and exploitation (player PD) with corresponding values of feasibility as shown in the following table. As in previous cases, there is always at least one F-N Nash equilibrium for all degrees of orness, but the higher the degree of orness the smaller the number of F-N Nash equilibria.

Table 18 Values of nuance and feasibility for trust game with SH cautious-spirited and PD mean-spirited players

SH: Cautious	ν-trust	FFF	VHH	HHH	MMM	LLL	VLL	NNN
	φ-feasibility	NNN	VLL	LLL	MMM	HHH	VHH	FFF
PD: Mean	ν-exploit	FFF	VHH	HHH	MMM	LLL	VLL	NNN
	φ-feasibility	FFF	VHH	HHH	MMM	LLL	VLL	NNN

Table 19 Highest Nash equilibria for ω=0.1 for trust game with SH cautious-spirited and PD mean-spirited players.

ω=0.1

Row	ν_1^k	φ_1^k	ν_2^l	φ_2^l	$\zeta(\nu_1^k)$	$\zeta(\varphi_1^k)$	$\zeta(\nu_2^l)$	$\zeta(\varphi_2^l)$	$\nu *^{pr}_{k_1^* l_2^*}$	$\varphi *^{pr}_{k_1^* l_2^*}$	Nash
1	FFF	NNN	FFF	FFF	FFF	MMM	VHH	MMM	MMM	MMM	F-N
2	FFF	NNN	MMM	MMM	FFF	MMM	MMM	LLL	MMM	MMM	F-N
3	FFF	NNN	VLL	VLL	FFF	MMM	VHH	MMM	MMM	MMM	F-N
4	FFF	NNN	NNN	NNN	FFF	MMM	FFF	MMM	MMM	MMM	F-N
5	VHH	VLL	FFF	FFF	VHH	MMM	VHH	MMM	MMM	MMM	F-N
6	VHH	VLL	MMM	MMM	VHH	MMM	MMM	LLL	MMM	MMM	F-N
7	VHH	VLL	VLL	VLL	VHH	MMM	VHH	MMM	MMM	MMM	F-N
8	VHH	VLL	NNN	NNN	VHH	MMM	FFF	MMM	MMM	MMM	F-N
9	MMM	MMM	FFF	FFF	MMM	LLL	VHH	MMM	MMM	MMM	F-N
10	MMM	MMM	VLL	VLL	MMM	LLL	VHH	MMM	MMM	MMM	F-N
11	MMM	MMM	NNN	NNN	MMM	LLL	FFF	MMM	MMM	MMM	F-N
12	NNN	FFF	FFF	FFF	HHH	MMM	VHH	MMM	MMM	MMM	F-N
13	NNN	FFF	MMM	MMM	HHH	MMM	MMM	LLL	MMM	MMM	F-N
14	NNN	FFF	VLL	VLL	HHH	MMM	VHH	MMM	MMM	MMM	F-N
15	NNN	FFF	NNN	NNN	HHH	MMM	FFF	MMM	MMM	MMM	F-N

Table 20 Highest Nash equilibria for ω=0.6 for trust game with SH cautious-spirited and PD mean-spirited players.

ω=0.6

Row	v_1^k	φ_1^k	v_2^l	φ_2^l	$\zeta(v_1^k)$	$\zeta(\varphi_1^k)$	$\zeta(v_2^l)$	$\zeta(\varphi_2^l)$	$v*_{k_1^*l_2^*}^{pr}$	$\varphi*_{k_1^*l_2^*}^{pr}$	Nash
1	FFF	NNN	FFF	FFF	FFF	HHH	HHH	MMM	MMM	HHH	F
2	FFF	NNN	LLL	LLL	FFF	HHH	VHH	MMM	MMM	HHH	F
3	FFF	NNN	VLL	VLL	FFF	HHH	VHH	MMM	MMM	HHH	F
4	FFF	NNN	NNN	NNN	FFF	HHH	FFF	HHH	HHH	HHH	F-N
5	VHH	VLL	NNN	NNN	VHH	MMM	FFF	HHH	MMM	HHH	F
6	HHH	LLL	NNN	NNN	HHH	MMM	FFF	HHH	MMM	HHH	F
7	NNN	FFF	NNN	NNN	HHH	MMM	FFF	HHH	MMM	HHH	F

Table 21 Highest Nash equilibria for ω=0.9 for trust game with SH cautious-spirited and PD mean-spirited players.

ω=0.9

Row	v_1^k	φ_1^k	v_2^l	φ_2^l	$\zeta(v_1^k)$	$\zeta(\varphi_1^k)$	$\zeta(v_2^l)$	$\zeta(\varphi_2^l)$	$v*_{k_1^*l_2^*}^{pr}$	$\varphi*_{k_1^*l_2^*}^{pr}$	Nash
1	FFF	NNN	FFF	FFF	FFF	FFF	VHH	VHH	VHH	FFF	F
2	FFF	NNN	VHH	VHH	FFF	FFF	VHH	VHH	VHH	FFF	F
3	FFF	NNN	HHH	HHH	FFF	FFF	HHH	HHH	HHH	FFF	F
4	FFF	NNN	MMM	MMM	FFF	FFF	MMM	MMM	MMM	FFF	F
5	FFF	NNN	LLL	LLL	FFF	FFF	HHH	HHH	HHH	FFF	F
6	FFF	NNN	VLL	VLL	FFF	FFF	VHH	VHH	VHH	FFF	F
7	FFF	NNN	NNN	NNN	FFF	FFF	FFF	FFF	FFF	FFF	F-N
8	VHH	VLL	NNN	NNN	VHH	VHH	FFF	FFF	VHH	FFF	F
9	HHH	LLL	NNN	NNN	HHH	HHH	FFF	FFF	HHH	FFF	F
10	MMM	MMM	NNN	NNN	MMM	MMM	FFF	FFF	MMM	FFF	F
11	LLL	HHH	NNN	NNN	HHH	HHH	FFF	FFF	HHH	FFF	F
12	VLL	VHH	NNN	NNN	HHH	VHH	FFF	FFF	VHH	FFF	F
13	NNN	FFF	NNN	NNN	HHH	VHH	FFF	FFF	VHH	FFF	F

For very low levels of orness **ω=0.1**, I find 15 F-N Nash equilibria, all at a moderate level of trust and all moderately feasible (Table 19). Among the 15 different pathways of reaching this type of equilibrium rows 1, 12, and 15 are especially interesting. In row 1 SH initially perceives full trust as not feasible and PD initially perceives full exploitation as fully feasible. In row 12 SH initially perceives no trust as fully feasible and PD initially perceives full exploitation as fully feasible. In row 15 SH initially perceives no trust as fully feasible and PD initially perceives no exploitation as not feasible. Yet, in all three cases playing the game with LFL produces a stable equilibrium even if it is only with a moderate trust which is moderately feasible.

For intermediate levels of orness **ω=0.6**, I find 7 Nash equilibria, one of which is of F-N type while all others are of F type (table 20). The F-N one emerges under conditions where SH initially perceives full trust as not feasible and PD initially perceives full exploitation as fully feasible. The outcome under LFL is F-N Nash equilibrium with high trust which is highly feasible. The F equilibria are also highly feasible but have a moderate level of trust.

For very high levels of orness ω=0.9, I find 13 Nash equilibria, one of which is F-N and the others are F type (Table 21). All equilibria have a high degree of feasibility. The F-N equilibrium has a full level of trust. The others are distributed in terms of the level of trust as follows: 6 with very high level of trust, 4 with high level of trust, and 2 with moderate level of trust.

SH naive-spirited and PD mean-spirited:

In this situation, we have a naïve-spirited SH player against a mean-spirited PD player. That is: player SH initially perceives full trust as fully feasible and no trust as not feasible. Player PD initially perceives full exploitation as fully feasible and no exploitation as not feasible. However, for any degree of orness, SH and PD are able to find a stable solution with at least a moderate level of trust and which is moderately feasible.

Table 22 Values of nuance and feasibility for trust game with SH naive-spirited and PD mean-spirited players

SH: Naive	ν-trust	FFF	VHH	HHH	MMM	LLL	VLL	NNN
	φ-feasibility	FFF	VHH	HHH	MMM	LLL	VLL	NNN
PD: Mean	ν-exploit	FFF	VHH	HHH	MMM	LLL	VLL	NNN
	φ-feasibility	FFF	VHH	HHH	MMM	LLL	VLL	NNN

For very low levels of orness **ω=0.1**, I find 16 F-N Nash equilibria, all of which at a moderate level of trust and all are moderately feasible (Table 23). Rows 1, 5, 12, and 16 are remarkable, however. In row 1 SH initially perceives full trust as fully feasible but PD perceives full exploitation as fully feasible. This is the equivalent of the crisp game of trust. Rational actors would end up not producing trust when playing according to Boolean logic. However, in the present game SH and PD are playing according to LFL and hence are able to produce a stable Nash equilibrium with a moderate level of trust and a moderate level of feasibility. Similar remarks apply to rows 3, 12, and 16.

For intermediate values of orness **ω=0.6**, I find 7 Nash equilibria, one of which is of F-N type and has a high level of trust (Table 24). The others are of F type and are at a moderate level of trust. All equilibria have a high degree of feasibility. The F-N Nash equilibrium emerges under a condition where SH initially perceives

Table 23 Highest Nash equilibria for ω=0.1 for trust game with SH naive-spirited and PD mean-spirited players.

ω=0.1

Row	v_1^k	φ_1^k	v_2^l	φ_2^l	$\zeta(v_1^k)$	$\zeta(\varphi_1^k)$	$\zeta(v_2^l)$	$\zeta(\varphi_2^l)$	$v*^{pr}_{k_1^* l_2^*}$	$\varphi*^{pr}_{k_1^* l_2^*}$	Nash
1	FFF	FFF	FFF	FFF	VHH	MMM	FFF	MMM	MMM	MMM	F-N
2	FFF	FFF	MMM	MMM	VHH	MMM	MMM	LLL	MMM	MMM	F-N
3	FFF	FFF	LLL	LLL	VHH	MMM	MMM	LLL	MMM	MMM	F-N
4	FFF	FFF	VLL	VLL	VHH	MMM	MMM	MMM	MMM	MMM	F-N
5	FFF	FFF	NNN	NNN	VHH	MMM	MMM	MMM	MMM	MMM	F-N
6	MMM	MMM	FFF	FFF	MMM	LLL	FFF	MMM	MMM	MMM	F-N
7	MMM	MMM	VLL	VLL	MMM	LLL	MMM	MMM	MMM	MMM	F-N
8	MMM	MMM	NNN	NNN	MMM	LLL	MMM	MMM	MMM	MMM	F-N
9	VLL	VLL	FFF	FFF	HHH	MMM	FFF	MMM	MMM	MMM	F-N
10	VLL	VLL	VLL	VLL	HHH	MMM	MMM	MMM	MMM	MMM	F-N
11	VLL	VLL	NNN	NNN	HHH	MMM	MMM	MMM	MMM	MMM	F-N
12	NNN	NNN	FFF	FFF	HHH	MMM	FFF	MMM	MMM	MMM	F-N
13	NNN	NNN	MMM	MMM	HHH	MMM	MMM	LLL	MMM	MMM	F-N
14	NNN	NNN	LLL	LLL	HHH	MMM	MMM	LLL	MMM	MMM	F-N
15	NNN	NNN	VLL	VLL	HHH	MMM	MMM	MMM	MMM	MMM	F-N
16	NNN	NNN	NNN	NNN	HHH	MMM	MMM	MMM	MMM	MMM	F-N

Table 24 Highest Nash equilibria for ω=0.6 for trust game with SH naive-spirited and PD mean-spirited players.

ω=0.6

Row	v_1^k	φ_1^k	v_2^l	φ_2^l	$\zeta(v_1^k)$	$\zeta(\varphi_1^k)$	$\zeta(v_2^l)$	$\zeta(\varphi_2^l)$	$v*^{pr}_{k_1^* l_2^*}$	$\varphi*^{pr}_{k_1^* l_2^*}$	Nash
1	FFF	FFF	NNN	NNN	HHH	MMM	MMM	HHH	MMM	HHH	F
2	LLL	LLL	NNN	NNN	MMM	MMM	MMM	HHH	MMM	HHH	F
3	VLL	VLL	NNN	NNN	HHH	MMM	MMM	HHH	MMM	HHH	F
4	NNN	NNN	FFF	FFF	HHH	HHH	HHH	MMM	MMM	HHH	F
5	NNN	NNN	LLL	LLL	HHH	HHH	MMM	MMM	MMM	HHH	F
6	NNN	NNN	VLL	VLL	HHH	HHH	MMM	MMM	MMM	HHH	F
7	NNN	NNN	NNN	NNN	HHH	HHH	MMM	HHH	HHH	HHH	F-N

no trust as not feasible and PD perceives no exploitation as not feasible. Yet, playing the game according to LFL allows the players to build a stable Nash equilibrium with a high level of trust which is highly feasible.

For very high levels of orness **ω=0.9**, I find 13 Nash equilibria, one of which is of F-N type at a full level of trust (Table 25). The other 12 equilibria are of F type

Table 25 Highest Nash equilibria for ω=0.9 for trust game with SH naive-spirited and PD mean-spirited players.

ω=0.9

Row	v_1^k	φ_1^k	v_2^l	φ_2^l	$\zeta(v_1^k)$	$\zeta(\varphi_1^k)$	$\zeta(v_2^l)$	$\zeta(\varphi_2^l)$	$v*^{pr}_{k_1^* l_2^*}$	$\varphi*^{pr}_{k_1^* l_2^*}$	Nash
1	FFF	FFF	NNN	NNN	VHH	VHH	MMM	FFF	VHH	FFF	F
2	VHH	VHH	NNN	NNN	VHH	VHH	MMM	FFF	VHH	FFF	F
3	HHH	HHH	NNN	NNN	HHH	HHH	MMM	FFF	HHH	FFF	F
4	MMM	MMM	NNN	NNN	MMM	MMM	MMM	FFF	MMM	FFF	F
5	LLL	LLL	NNN	NNN	HHH	HHH	MMM	FFF	HHH	FFF	F
6	VLL	VLL	NNN	NNN	HHH	VHH	MMM	FFF	VHH	FFF	F
7	NNN	NNN	FFF	FFF	HHH	FFF	FFF	VHH	VHH	FFF	F
8	NNN	NNN	VHH	VHH	HHH	FFF	VHH	VHH	VHH	FFF	F
9	NNN	NNN	HHH	HHH	HHH	FFF	HHH	HHH	HHH	FFF	F
10	NNN	NNN	MMM	MMM	HHH	FFF	MMM	MMM	MMM	FFF	F
11	NNN	NNN	LLL	LLL	HHH	FFF	MMM	HHH	HHH	FFF	F
12	NNN	NNN	VLL	VLL	HHH	FFF	MMM	VHH	VHH	FFF	F
13	NNN	NNN	NNN	NNN	HHH	FFF	MMM	FFF	FFF	FFF	F-N

at a various levels of trust. All 13 equilibria are fully feasible. What is more remarkable is that the F-N Nash equilibrium emerges in a situation where SH initially believes that no trust is not feasible and PD initially believes that no exploitation is not feasible. The LFL rules of the game allow SH and PD to arrive at a stable Nash equilibrium at a full level of trust which is fully feasible.

SH naive-spirited and PD nice-spirited:

In this situation, we have a naïve-spirited SH player against a nice-spirited PD player. That is: player SH initially perceives full trust as fully feasible, and no trust as not feasible. Player PD initially perceives full exploitation as not feasible, and no exploitation as fully feasible. However, for any degree of orness, SH and PD are able to find a stable solution with at least a moderate level of trust and which is moderately feasible.

Table 26 Values of nuance and feasibility for trust game with SH naive-spirited and PD nice-spirited players

SH: Naïve	ν-trust	FFF	VHH	HHH	MMM	LLL	VLL	NNN
	φ-feasibility	FFF	VHH	HHH	MMM	LLL	VLL	NNN
PD: Nice	ν-exploit	FFF	VHH	HHH	MMM	LLL	VLL	NNN
	φ-feasibility	NNN	VLL	LLL	MMM	HHH	VHH	FFF

Table 27 Highest Nash equilibria for ω=0.1 for trust game with SH naive-spirited and PD nice-spirited players.

ω=0.1

Row	v_1^k	φ_1^k	v_2^l	φ_2^l	$\zeta(v_1^k)$	$\zeta(\varphi_1^k)$	$\zeta(v_2^l)$	$\zeta(\varphi_2^l)$	$v^{*pr}_{k_1^* l_2^*}$	$\varphi^{*pr}_{k_1^* l_2^*}$	Nash
1	FFF	FFF	FFF	NNN	FFF	MMM	FFF	MMM	MMM	MMM	F-N
2	FFF	FFF	VHH	VLL	FFF	MMM	VHH	MMM	MMM	MMM	F-N
3	FFF	FFF	MMM	MMM	FFF	MMM	MMM	LLL	MMM	MMM	F-N
4	FFF	FFF	LLL	HHH	FFF	MMM	MMM	LLL	MMM	MMM	F-N
5	FFF	FFF	VLL	VHH	FFF	MMM	LLL	LLL	MMM	MMM	F-N
6	FFF	FFF	NNN	FFF	FFF	MMM	LLL	MMM	MMM	MMM	F-N
7	MMM	MMM	FFF	NNN	MMM	LLL	FFF	MMM	MMM	MMM	F-N
8	MMM	MMM	VHH	VLL	MMM	LLL	VHH	MMM	MMM	MMM	F-N
9	MMM	MMM	NNN	FFF	MMM	LLL	LLL	MMM	MMM	MMM	F-N
10	VLL	VLL	FFF	NNN	VHH	MMM	FFF	MMM	MMM	MMM	F-N
11	VLL	VLL	VHH	VLL	VHH	MMM	VHH	MMM	MMM	MMM	F-N
12	VLL	VLL	MMM	MMM	VHH	MMM	MMM	LLL	MMM	MMM	F-N
13	VLL	VLL	LLL	HHH	VHH	MMM	MMM	LLL	MMM	MMM	F-N
14	VLL	VLL	VLL	VHH	VHH	MMM	LLL	LLL	MMM	MMM	F-N
15	VLL	VLL	NNN	FFF	VHH	MMM	LLL	MMM	MMM	MMM	F-N
16	NNN	NNN	FFF	NNN	FFF	MMM	FFF	MMM	MMM	MMM	F-N
17	NNN	NNN	VHH	VLL	FFF	MMM	VHH	MMM	MMM	MMM	F-N
18	NNN	NNN	MMM	MMM	FFF	MMM	MMM	LLL	MMM	MMM	F-N
19	NNN	NNN	LLL	HHH	FFF	MMM	MMM	LLL	MMM	MMM	F-N
20	NNN	NNN	VLL	VHH	FFF	MMM	LLL	LLL	MMM	MMM	F-N
21	NNN	NNN	NNN	FFF	FFF	MMM	LLL	MMM	MMM	MMM	F-N

For very low levels of orness **ω=0.1**, I find 21 F-N Nash equilibria, all of which are at a moderate level of trust and all are moderately feasible (Table 27). Rows 1, 6, 16, and 21 are remarkable, however. In row 1 SH initially perceives full trust as fully feasible and PD perceives full exploitation as not feasible. This is a departure from the crisp equivalent form of the trust game. Rational actors would end up not producing trust when playing according to Boolean logic. However, in the present game SH and PD are playing according to LFL and hence are able to produce a stable Nash equilibrium with a moderate level of trust and a moderate level of feasibility. Similar remarks apply to rows 6, 16, and 21.

For intermediate values of orness **ω=0.6**, I find 8 Nash equilibria, two of which are of F-N type and have a high level of trust (Table 28). The others are of F type and are at a moderate level of trust. All equilibria have a high degree of feasibility. The F-N Nash equilibria emerge under conditions where (row 4) SH initially perceives no trust as not feasible and PD perceives exploitation as not feasible and

Table 28 Highest Nash equilibria for ω=0.6 for trust game with SH naive-spirited and PD nice-spirited players.

ω=0.6

Row	v_1^k	φ_1^k	v_2^l	φ_2^l	$\zeta(v_1^k)$	$\zeta(\varphi_1^k)$	$\zeta(v_2^l)$	$\zeta(\varphi_2^l)$	$v^{*pr}_{k_1^* l_2^*}$	$\varphi^{*pr}_{k_1^* l_2^*}$	Nash
1	FFF	FFF	FFF	NNN	HHH	MMM	FFF	HHH	MMM	HHH	F
2	LLL	LLL	FFF	NNN	VHH	MMM	FFF	HHH	MMM	HHH	F
3	VLL	VLL	FFF	NNN	VHH	MMM	FFF	HHH	MMM	HHH	F
4	NNN	NNN	FFF	NNN	FFF	HHH	FFF	HHH	HHH	HHH	F-N
5	NNN	NNN	VHH	VLL	FFF	HHH	VHH	MMM	MMM	HHH	F
6	NNN	NNN	HHH	LLL	FFF	HHH	HHH	MMM	MMM	HHH	F
7	NNN	NNN	VLL	VHH	FFF	HHH	LLL	LLL	HHH	MMM	N
8	NNN	NNN	NNN	FFF	FFF	HHH	LLL	MMM	HHH	HHH	F-N

Table 29 Highest Nash equilibria for ω=0.9 for trust game with SH naive-spirited and PD nice-spirited players.

ω=0.9

Row	v_1^k	φ_1^k	v_2^l	φ_2^l	$\zeta(v_1^k)$	$\zeta(\varphi_1^k)$	$\zeta(v_2^l)$	$\zeta(\varphi_2^l)$	$v^{*pr}_{k_1^* l_2^*}$	$\varphi^{*pr}_{k_1^* l_2^*}$	Nash
1	FFF	FFF	FFF	NNN	FFF	VHH	FFF	FFF	VHH	FFF	F
2	VHH	VHH	FFF	NNN	VHH	VHH	FFF	FFF	VHH	FFF	F
3	HHH	HHH	FFF	NNN	HHH	HHH	FFF	FFF	HHH	FFF	F
4	MMM	MMM	FFF	NNN	MMM	MMM	FFF	FFF	MMM	FFF	F
5	LLL	LLL	FFF	NNN	HHH	HHH	FFF	FFF	HHH	FFF	F
6	VLL	VLL	FFF	NNN	VHH	VHH	FFF	FFF	VHH	FFF	F
7	NNN	NNN	FFF	NNN	FFF	FFF	FFF	FFF	FFF	FFF	F-N
8	NNN	NNN	VHH	VLL	FFF	FFF	VHH	VHH	VHH	FFF	F
9	NNN	NNN	HHH	LLL	FFF	FFF	HHH	HHH	HHH	FFF	F
10	NNN	NNN	MMM	MMM	FFF	FFF	MMM	MMM	MMM	FFF	F
11	NNN	NNN	LLL	HHH	FFF	FFF	MMM	HHH	HHH	FFF	F
12	NNN	NNN	VLL	VHH	FFF	FFF	LLL	VHH	VHH	FFF	F
13	NNN	NNN	NNN	FFF	FFF	FFF	LLL	VHH	VHH	FFF	F

(row 8) SH initially perceives no trust as not feasible and PD perceives no exploitation as fully feasible. Playing the game according to LFL allows the players to build a stable Nash equilibrium with a high level of trust which is highly feasible.

For very high levels of orness **ω=0.9** (Table 29), I find 13 Nash equilibria, one of which is of F-N type at a full level of trust (row7). The other 12 equilibria are of F type at a various levels of trust. All 13 equilibria are fully feasible. The F-N Nash equilibrium emerges in a situation where SH initially believes that no trust is not feasible but PD initially believes that full exploitation is not feasible. The LFL rules of the game allow SH and PD to arrive at a stable Nash equilibrium at a full level of trust which is fully feasible.

In sum, in the LFL game of trust, players are always able to build some (at least moderate) level trust with some (at least moderate) degree of feasibility. For all values of orness, there exists at least one Nash equilibrium of F-N type. Playing the trust game according to LFL rules allows the players to escape the trust dilemma that would normally emerge in a crisp game of trust without communication or punishment incentives/disincentives.

Chapter 5
Linguistic Fuzzy-Logic Social Game

In Chapter 4, I developed a new game-theoretic approach based, not on conventional Boolean two-valued logic, but instead on linguistic fuzzy logic which admits linguistic truth values. A linguistic fuzzy game is defined with linguistic fuzzy strategies, linguistic fuzzy preferences, and the rules of reasoning and inferences of the game operate according to linguistic fuzzy logic, not Boolean logic. This leads to the introduction of a new notion of fuzzy domination and Nash equilibrium which are based not on the usual "greater than" relation ordering but rather on a more general form of relation termed linguistic fuzzy relation. Each agent models others as linguistic fuzzy rational agents and tries to find a linguistic fuzzy Nash equilibrium that will achieve the highest linguistic fuzzy payoff.

In a linguistic fuzzy N-person game, each agent models others as linguistic fuzzy rational agents and tries to find a linguistic fuzzy Nash equilibrium that will achieve the highest linguistic fuzzy preference defined according to a linguistic fuzzy logic. Conversely, if the linguistic fuzzy relations and all other components of the game structure are simplified into a crisp two-valued logic, the linguistic fuzzy game reduces to the conventional game. As such, conventional game theory can be viewed as a limiting case of the linguistic fuzzy game. In this chapter I apply the new approach to a social game of cooperation with N players. I find that there is always an optimum strong Nash equilibrium which is Pareto optimal, thereby lifting many of the dilemmas that emerge in crisp game theory in social games.

I generalize the formalism from a 2-player game of the previous chapter to a social game. I consider a social game of cooperation among N actors with a Prisoner's Dilemma game as the stage game. This social game has typically been used to analyze the dilemmas of collective action and the problem of opportunism in political science and economic literature.[1] The formalism of this section can be applied to other social games using different stage games. The model is based on a finite population of players, which are drawn at random within a period to play a

[1] Conventional game-theoretic literature argues that collective cooperation emerges when individual participants do not discount the future too much and face some sort of punishment – that is, a loss of future payoffs for non-cooperation. In social situations wherein the actors heavily discount future relations and do not face the specter of punishment for non-cooperation, it is believed that opportunism, which is ubiquitous in social relations and can arise at the level of both individual and inter-group interaction, would impede the emergence and/or consolidation of collective cooperation. See, for illustrative purposes, Fearon and Laitin (1996); Arfi (2000).

B. Arfi: Linguistic Fuzzy Logic Methods in Social Sciences, STUDFUZZ 253, pp. 105–119.
springerlink.com

stage game and adjust their behavior over time accordingly. I consider two different scenarios: (1) dyad-based game where the players play the game in pairs and (2) cluster-based game where the players play the game in clusters containing N_{nn} nearest neighbors. The dyad-based game can be used, for example, to model the recent US approach to the International Criminal Court. The US strategy is to sign bilateral treaties with all adherents to International Criminal Court as a condition of the US adherence to the treaty, with the bilateral agreements granting immunity to US troops and other government officials who might potentially fall under the jurisdiction of the ICC treaty. The principal US objection is that the ICC provisions could enable states hostile to the US to abuse the treaty mandate and turn it against US military personnel who are engaged in overseas missions (Bolton, 2001). One could well imagine other world powers emulating the US and thus playing the same type of stage game as a condition for adhering to the treaty. The cluster-based game can be considered, for example, as a simplified stylization of the role that caucuses play in the US Congress at those times when the Congress is engaged in full internal negotiations. The cluster-based game can also be used to model the problem of governance in multiethnic polities wherein the internal dynamics of the various ethnic communities cannot be ignored due to dilemmas of collective action within the ethnic groups themselves (Arfi, 2000).

1 Dyad-Based Social PD Game

With N players we can have N(N-1)/2 pairs of players. We assume that every pair plays a LFL-PD game. The profile of the social game is a conjunction of the profiles of all N(N-1)/2 dyads. This assumes that the dyad-profiles are independent of one another. Later in the chapter we explore how this assumption can be relaxed into cluster-profiles with a cluster consisting of a small number of players. The conjunction of these cluster profiles then constitutes a profile for the social game. Using the same notation as in the previous chapter, a profile for a dyad (i,j) of players $i \neq j$ in the LFL-Game is given by:

$$GP\left(i,l_i;j,n_j\right)=\Re_{l_i\left(k_j\right)}\wedge\Re_{\left(m_i\right)n_j}=LWC\left\{\Re_{l_i\left(k_j\right)};\Re_{\left(m_i\right)n_j}\right\} \tag{80}$$

Where:

$$\Re_{l_i\ \left(k_j\right)}=\Re_{l_i\ k_1}\vee\Re_{l_i\ k_2}\vee\cdots\vee\Re_{l_i\ k_{qj}}=LWD\left\{\Re_{l_i\ k_1};\Re_{l_i\ k_2};\cdots;\Re_{l_i\ k_{qj}}\right\} \tag{81}$$

$$\Re_{\left(m_i\right)n_j}=\Re_{m_1\ n_j}\vee\Re_{m_2\ n_j}\vee\cdots\vee\Re_{m_{qi}\ n_j}=LWD\left\{\Re_{m_1\ n_j};\Re_{m_2\ n_j};\cdots;\Re_{m_{qi}\ n_j}\right\} \tag{82}$$

The profile for player i playing the dyad-based social game, that is, the game with all other N-1 players $j \in \{1,...,i-1,i+1,N\}$, each of whom with q_j LFL-choices, is:

$$SGP\left[i,l_i\right] = \bigwedge_{j=i+1,...,N} \left\{ \bigvee_{n_j=1,...,q_j} GP\left(i,l_i;\, j,n_j\right) \right\} \tag{83}$$

Note that the conjunction goes from $i+1$ to N only to avoid counting the same pair of players twice. That is:

$$SGP\left[i,l_i\right] = \underset{j=i+1,...,N}{LWC} \left\{ \underset{n_j=1,...,q_j}{LWD} \left\{ GP\left(i,l_i;\, j,n_j\right) \right\} \right\} \tag{84}$$

The profile for the N-player social game with dyad-based PD stage game, with $l \in \{1,..., q_l\}$ LFL-choices, is then given by:

$$SGP\left[N,l\right] = \bigwedge_{i=1,...,N-1} \left\{ SGP\left[i,l\right] \right\} = LWC\left\{ SGP\left[i,l\right] \middle| i=1,...,N-1 \right\} \tag{85}$$

I assume that the players engaged in the stage games are all of one of two types: nice or mean, with their initial perceptions on cooperation as shown in Table 1. This produces a society of nice actors and a society of mean actors. The question pursued is to see whether one of these two societies is more capable of producing more societal cooperation, that is, at what level of cooperation and at what level of feasibility.

Table 1 Values of nuance and feasibility for one nice- and mean-spirited player.

PD: Nice	ν-cooperation	FFF	VHH	HHH	MMM	LLL	VLL	NNN
	φ-feasibility	FFF	VHH	HHH	MMM	LLL	VLL	NNN
PD: Mean	ν-cooperation	FFF	VHH	HHH	MMM	LLL	VLL	NNN
	φ-feasibility	NNN	VLL	LLL	MMM	HHH	VHH	FFF

I assume that the social game has 200 participants and then explore the emergence of equilibrium for very low, intermediate, and very high values of orness. In Tables 2, 3, and 4, I compare the solutions for a society of nice players to a society of mean players at the same level of orness and with the same population size.

For very low levels of orness (Table 2), the two societies fare in a similar manner: they both produce an F-N Nash equilibrium with full cooperation at a moderate level of feasibility. In addition, each produces two F Nash equilibria, one of which is at a moderate level of cooperation with a moderate level of feasibility for both societies. The two societies however differ on the second F equilibrium.

Table 2 Comparing possible Nash social equilibria for ω=0.1 for dyad-based PD social game in society of N nice-spirited players to a society of N mean-spirited players.

ω=0.1; N =200									
Stage Game: PD Mean					Stage Game: PD Nice				
v_1^k	φ_1^k	v^{*SG}	φ^{*SG}	Nash	v_1^k	φ_1^k	v^{*SG}	φ^{*SG}	Nash
FFF	NNN	FFF	MMM	F-N	FFF	FFF	FFF	MMM	F-N
VHH	VLL	VHH	MMM	F	VHH	VHH	VHH	LLL	
HHH	LLL	HHH	LLL		HHH	HHH	HHH	LLL	
MMM	MMM	MMM	LLL		MMM	MMM	MMM	LLL	
LLL	HHH	MMM	LLL		LLL	LLL	MMM	LLL	
VLL	VHH	LLL	LLL		VLL	VLL	MMM	MMM	F
NNN	FFF	MMM	MMM	F	NNN	NNN	MMM	MMM	F

Whereas in the society of nice players the second F equilibrium is at a moderate level of cooperation, in the society of mean players it is at a very high level of cooperation. In this case, the society of mean players counter-intuitively fares much better than a society of nice players in building cooperation!

At intermediate values of orness (Table 3), the mean-populated society produces a F-N Nash equilibrium with full cooperation at a high level of feasibility. The nice-populated society is able to produce one F and one N equilibrium, but no F-N equilibrium. None of these two equilibria is at a level of full cooperation. The society of mean players fares much better than one of nice players in resolving the prisoner's dilemma at a societal level when the game is played with LFL rules.

For very high levels of orness (Table 4), there is some sort of convergence between the two societies as both produce one F-N Nash equilibrium at a full level of cooperation which is fully feasible. The nice-populated society has an

Table 3 Comparing possible Nash social equilibria for ω=0.5 for social game in society of N nice-spirited players to a society of N mean-spirited players.

ω=0.5; N =200									
Stage Game: PD Mean					Stage Game: PD Nice				
v_1^k	φ_1^k	v^{*SG}	φ^{*SG}	Nash	v_1^k	φ_1^k	v^{*SG}	φ^{*SG}	Nash
FFF	NNN	FFF	HHH	F-N	FFF	FFF	VHH	LLL	N
VHH	VLL	VHH	MMM		VHH	VHH	HHH	LLL	
HHH	LLL	VHH	LLL		HHH	HHH	HHH	LLL	
MMM	MMM	HHH	LLL		MMM	MMM	MMM	LLL	
LLL	HHH	HHH	LLL		LLL	LLL	MMM	LLL	
VLL	VHH	MMM	LLL		VLL	VLL	MMM	MMM	
NNN	FFF	MMM	LLL		NNN	NNN	HHH	HHH	F

Table 4 Comparing possible Nash social equilibria for ω=0.9 for social game in society of N nice-spirited players to a society of N mean-spirited players.

ω=0.9; N =200									
Stage Game: PD Mean					Stage Game: PD Nice				
v_1^k	φ_1^k	v^{*SG}	φ^{*SG}	Nash	v_1^k	φ_1^k	v^{*SG}	φ^{*SG}	Nash
FFF	NNN	FFF	FFF	F-N	FFF	FFF	FFF	VHH	N
VHH	VLL	VHH	VHH		VHH	VHH	VHH	VHH	
HHH	LLL	HHH	HHH		HHH	HHH	HHH	HHH	
MMM	MMM	MMM	MMM		MMM	MMM	MMM	MMM	
LLL	HHH	HHH	HHH		LLL	LLL	HHH	HHH	
VLL	VHH	VHH	VHH		VLL	VLL	VHH	VHH	
NNN	FFF	VHH	VHH		NNN	NNN	FFF	FFF	F-N

additional equilibrium at full cooperation with a very high level of feasibility. Therefore, the nice-populated society fares better than the mean-populated one for very high levels of orness. Niceness is thus relatively speaking punished at low values of orness but is positively rewarded at high values of orness.

2 Crisp Cluster-Based Social PD Game

We assume that each player is engaged in a PD-like game within a cluster of N_{nn} nearest-neighbor players. In other words, we assume a PD stage game with N_{nn} players. In order to develop the formalism for this case as transparently as possible let us first consider the crisp version of the game. For the sake of illustration, consider a case where $N_{nn} = 5$. A starting point is to express the rules of inference of the PD stage game with 5 players, P1,…, P5. As in the dyadic PD game we still assume that each player has initially two choices, C and D.

In the 2-player game, each player has a possibility of 2x2=4 rules of inferences. For P1, we have the following set of rules $\left\{\mathfrak{R}_{11}, \mathfrak{R}_{12}, \mathfrak{R}_{21}, \mathfrak{R}_{22}\right\}$, as discussed in the previous chapter. For a PD game with 5 players, each player would have 2^5=64 possible rules of inference. However, in the crisp PD game each player has only two strategic arrangements – cooperate or defect. If we denote by $\mathfrak{R}_{ij}^{mn}$ the rule combining the choices *i* = (*C* or *D*) and *j* = (*C* or *D*) of players *m* or *n*, respectively, we can write the possible strategic arrangements $SA[\cdot]$ for player 1 in the 5-player game as:

$$SA[1,C] = \left(\mathfrak{R}_{CC}^{12} \vee \mathfrak{R}_{CD}^{12}\right) \wedge \left(\mathfrak{R}_{CC}^{13} \vee \mathfrak{R}_{CD}^{13}\right) \wedge \left(\mathfrak{R}_{CC}^{14} \vee \mathfrak{R}_{CD}^{14}\right) \wedge \left(\mathfrak{R}_{CC}^{15} \vee \mathfrak{R}_{CD}^{15}\right) \tag{86}$$

$$SA[1,D]=\left(\Re_{DC}^{12}\vee\Re_{DD}^{12}\right)\wedge\left(\Re_{DC}^{13}\vee\Re_{DD}^{13}\right)\wedge\left(\Re_{DC}^{14}\vee\Re_{DD}^{14}\right)\wedge\left(\Re_{DC}^{15}\vee\Re_{DD}^{15}\right) \tag{87}$$

For a cluster-based PD game we have two typical profiles for a cluster with five players:

$$\Gamma(C)=\bigwedge_{i=1,\dots,5}SA[i,C]=\bigwedge_{i=1,\dots,4}\left\{\bigwedge_{j=i+1,\dots,5}\left(\Re_{CC}^{ij}\vee\Re_{CD}^{ij}\right)\right\} \tag{88}$$

$$\Gamma(D)=\bigwedge_{i=1,\dots,5}SA[i,D]=\bigwedge_{i=1,\dots,4}\left\{\bigwedge_{j=i+1,\dots,5}\left(\Re_{DC}^{ij}\vee\Re_{DD}^{ij}\right)\right\} \tag{89}$$

Let us generalize this to a game with more than two basic choices. We would have for player m $q_m=q$ initial choices from which the players can build different strategic arrangements in the game:[2]

$$SA[m,j]=\bigwedge_{n=1,\dots,m-1,m+1,\dots,N_{nn}}\left\{\bigvee_{k=1,\dots,q}\Re_{jk}^{mn}\right\} \tag{90}$$

Where:

$$\left\{\Re_{jk}^{mn}=\left[\left(s_j^m\wedge s_k^n\right)\rightarrow R_j\left(s_j^m;s_k^n\right)\right]\,\middle|\,j\in\{1,\dots,q\};k\in\{1,\dots,q\};m\in\{1,\dots,N_{nn}\}\neq n\in\{1,\dots,N_{nn}\}\right\} \tag{91}$$

Thus, more explicitly:

$$SA\left[m,j\right]=\bigwedge_{n=1,\dots m-1,m+1,\dots,N_{nn}}\left\{\bigvee_{k=1,\dots,q}\left\{\left(s_j^m\wedge s_k^n\right)\rightarrow R_j\left(s_j^m;s_k^n\right)\right\}\right\} \tag{92}$$

A typical game profile of a cluster with N_{nn} players is given by:

$$\Gamma_j\left(N_{nn},j\in\{1,\dots,q\}\right)=\bigwedge_{m=1,\dots,N_{nn}}SA\left[m,j\right] \tag{93}$$

That is:

$$\Gamma_j=\bigwedge_{m=1,\dots,N_{nn}-1}\left\{\bigwedge_{n=m+1,\dots,N_{nn}}\left\{\bigvee_{k=1,\dots,q}\left\{\left(s_j^m\wedge s_k^n\right)\rightarrow R_j\left(s_j^m;s_k^n\right)\right\}\right\}\right\} \tag{94}$$

---

[2] I am assuming for the sake of simplicity that all players have the same number q of initial choices.

The first conjunction runs from $m = 1$ to $m = N_{nn} - 1$ only and the third conjunction runs from $n = m + 1$ to $n = N_{nn}$ to avoid counting twice any pair of players.

Let us assume that the society is made up of $\aleph_c$ clusters, each having the same number of players N_{nn}. The clusters interact pairwise with one another through a 2x2 PD stage game. The task is to find whether there is a social equilibrium and its type, when it exists, depending on the number of players in a cluster N_{nn} and the number of clusters $\aleph_c$ in the society. The profile for two clusters α and β interacting through a PD game is:

$$CPA(\alpha, k; \beta, l; \aleph_c) = \Gamma_k^\alpha \wedge \Gamma_l^\beta \rightarrow R_k\left(\Gamma_k^\alpha; \Gamma_l^\beta\right) \tag{95}$$

The profile for cluster α playing the dyad-based PD game against $\aleph_c - 1$ clusters, that is, the game with clusters $\beta \in \{1, \ldots, \alpha - 1, \alpha + 1, \ldots, \aleph_c\}$, is:

$$CP(\alpha, k; N_{nn}; \aleph_c) = \bigwedge_{\beta = 1, \ldots, \alpha - 1, \alpha + 1, \ldots, N_{nn}} \left\{ \bigvee_{l = 1, \ldots, q} \left\{ \Gamma_k^\alpha \wedge \Gamma_l^\beta \rightarrow R_k\left(\Gamma_k^\alpha; \Gamma_l^\beta\right) \right\} \right\} \tag{96}$$

A societal profile for a society with $\aleph_c$ clusters, each made up of N_{nn} players, reads:

$$SP(\aleph_c; N_{nn}; k) = \bigwedge_{\alpha = 1, \ldots, \aleph_c} CP(\alpha, k; N_{nn}, \aleph_c) \tag{97}$$

That is:

$$SP(\aleph_c; N_{nn}; k) = \bigwedge_{\alpha = 1, \ldots, \aleph_c - 1} \left\{ \bigwedge_{\beta = \alpha + 1, \ldots, \aleph_c} \left\{ \bigvee_{l = 1, \ldots, q} \left\{ \Gamma_k^\beta \wedge \Gamma_l^\beta \rightarrow R_k\left(\Gamma_k^\alpha; \Gamma_l^\beta\right) \right\} \right\} \right\} \tag{98}$$

3 Fuzzy Cluster-Based Social PD Game

Let us now generalize this to a LFL-Game with $\aleph_c$ clusters. In the societal game the game profile is the result of two steps: first individual players facing one another within each cluster and second aggregated strategies for clusters against one another. Using the terminology of nuance and feasibility as in the previous games, the intra-cluster process would consist, first, in evaluating the degrees of nuance and feasibility for each cluster and, second, in evaluating the degrees of nuance and feasibility for a society made up of these clusters. The players are not allowed to interact individually across clusters. I assume that all clusters have the same number of players N_{nn}. Within a cluster α of N_{nn} players a typical strategic arrangement for player *m* with an initial choice s_j^m would now be:

$$SA[\alpha;m,j] = \underset{n=1,\ldots m-1,m+1,\ldots,N_{nn}}{LWC}\left\{\underset{k=1,\ldots,q}{LWD}\left\{LI\left\{LWC\left\{s_j^{\alpha,m};s_k^{\alpha,n}\right\};R_j\left(s_j^{\alpha,m};s_k^{\alpha,n}\right)\right\}\right\}\right\} \tag{99}$$

A typical LFL-profile for a cluster is:

$$\Gamma_j^\alpha = \underset{m=1,\ldots,N_{nn}}{LWC}\left\{SA[\alpha;m,j]\right\} \tag{100}$$

More explicitly:

$$\Gamma_j^\alpha = \underset{m=1,\ldots,N_{nn}-1}{LWC}\left\{\underset{n=m+1,\ldots,N_{nn}}{LWC}\left\{\underset{k=1,\ldots,q}{LWD}\left\{LI\left\{LWC\left\{s_j^{\alpha,m};s_k^{\alpha,n}\right\};R_j\left(s_j^{\alpha,m};s_k^{\alpha,n}\right)\right\}\right\}\right\}\right\} \tag{101}$$

The profile for cluster α playing the dyad-based PD game against $\aleph_c - 1$ clusters, that is, the game with clusters $\beta \in \{1,\ldots,\alpha-1,\alpha+1,\ldots,\aleph_c\}$, is:

$$CP(\alpha,j;N_{nn};\aleph_c) = \underset{\beta=1,\ldots,\alpha-1,\alpha+1,\ldots,N_{nn}}{LWC}\left\{\underset{l=1,\ldots,q}{LWD}\left\{LI\left\{LWC\left\{\Gamma_j^\alpha;\Gamma_l^\beta\right\};R_j\left(\Gamma_j^\alpha;\Gamma_l^\beta\right)\right\}\right\}\right\} \tag{102}$$

A societal profile for a society with $\aleph_c$ clusters, each made up of N_{nn} players, reads:

$$SP(\aleph_c;N_{nn};j) = \bigwedge_{\alpha=1,\ldots,\aleph_c} CP(\alpha,j;N_{nn},\aleph_c) = \underset{\alpha=1,\ldots,\aleph_c}{LWC}\left\{CP(\alpha,j;N_{nn},\aleph_c)\right\} \tag{103}$$

That is:

$$SP(\aleph_c;N_{nn};j) = \underset{\alpha=1,\ldots,\aleph_c-1}{LWC}\left\{\underset{\beta=\alpha+1,\ldots,N_{nn}}{LWC}\left\{\underset{l=1,\ldots,q}{LWD}\left\{LI\left\{LWC\left\{\Gamma_j^\alpha;\Gamma_l^\beta\right\};R_j\left(\Gamma_j^\alpha;\Gamma_l^\beta\right)\right\}\right\}\right\}\right\} \tag{104}$$

There are various possibilities of how to use the LFL-PD stage game combined with the type of players, that is, various levels of nice-spiritedness and mean-spiritedness. The latter can be considered as attributes of individuals within the same cluster as well as attributes of clusters themselves.

The degrees of nuance and feasibility are given for a strategic arrangement *SA* of a player *m* within cluster α by:

$$\left(v_{SA,j}^{\alpha,m};\varphi_{SA,j}^{\alpha,m}\right)=\underset{n=1,\ldots m-1,m+1,\ldots,N_{nn}}{LWC}\left\{\underset{k=1,\ldots,q}{LWD}\left\{LI\left\{LWC\left\{\left(v_{j}^{\alpha,m};\varphi_{j}^{\alpha,m}\right);\left(v_{k}^{\alpha,n};\varphi_{k}^{\alpha,n}\right)\right\};\left(v_{R_j}^{\alpha,m};\varphi_{R_j}^{\alpha,m}\right)\right\}\right\}\right\} \quad (105)$$

Where LWC and LWD have been defined in Chapter 2 in terms of the operators: MAX, MIN, and LOWA. This gives the following degrees of nuance and feasibility:

$$\left(v_{\Gamma}^{\alpha,j};\varphi_{\Gamma}^{\alpha,j}\right)=\underset{m=1,\ldots,N_{nn}-1}{LWC}\left\{\underset{n=m+1,\ldots,N_{nn}}{LWC}\left\{\underset{k=1,\ldots,q}{LWD}\left\{LI\left\{LWC\left\{\left(v_{j}^{\alpha,m};\varphi_{j}^{\alpha,m}\right);\left(v_{k}^{\alpha,n};\varphi_{k}^{\alpha,n}\right)\right\};\left(v_{R_j}^{\alpha,m};\varphi_{R_j}^{\alpha,m}\right)\right\}\right\}\right\}\right\} \quad (106)$$

A societal profile for a society with $\aleph_c$ clusters, each made up of N_{nn} players, reads:

$$\left(v_{SP}^{j};\varphi_{SP}^{j}\right)=\underset{\alpha=1,\ldots,\aleph_c-1}{LWC}\left\{\underset{\beta=\alpha+1,\ldots,\aleph_c}{LWC}\left\{\underset{l=1,\ldots,q}{LWD}\left\{LI\left\{LWC\left\{\left(v_{\Gamma}^{\alpha,j};\varphi_{\Gamma}^{\alpha,j}\right);\left(v_{\Gamma}^{\beta,l};\varphi_{\Gamma}^{\beta,l}\right)\right\};\left(v_{R_j}^{\alpha,j};\varphi_{R_j}^{\alpha,j}\right)\right\}\right\}\right\}\right\} \quad (107)$$

I explore one illustrative case where the basic stage game within a cluster is a PD one and the cluster-cluster stage game is also a PD game. The players within clusters can be one of two fuzzy types: more or less mean-spirited and more or less nice-spirited. The matrix of linguistic rankings is what determines the nature of a fuzzy game. I thus use as inputs in the internal game (that is, within clusters) and external game (between clusters) the matrices introduced in Chapter 4 when exploring the 2-player PD. The results for the game are grouped in Tables 5, 6, and 7. I display the results in terms of the type of players in the internal game, the degrees of orness in the internal and external games, and the number of players in the internal game and the number of clusters in the social game.

A remarkable result is that the number of clusters does not seem to play an important role in determining the final results as long as we have more than 3 clusters in the societal game. Thus, going, for example, from 10 to 100 clusters does not affect the final results. The game within each cluster is somewhat more sensitive to the number of players in a cluster. Nonetheless, this sensitiveness fades away as soon as the number of players reaches 5 or 6 players in a cluster. Hence, going from 10 to 100 players in a cluster does not change the degrees of nuance and feasibility of a cluster game.

It is useful to remember at this juncture two points. First, orness is defined for the degrees of feasibility only (see Chapters 2 and 4) and does not directly affect the aggregation process of the degrees of nuance. I use the term "directly" because to compute the degrees of nuance inescapably involves the degree of feasibility. Thus, we can say that orness affects the degree of nuance in a "second level" type

Table 5 Nash social equilibria for cluster-based social game with two nice-spirited players playing internal PD game.

$\omega_{cluster}$ = 0.05; $\omega_{Society}$ = 0.05; N_{nn} = 10; $\aleph_c$ = 10								
v_{PD}^1-coop	φ_{PD}^1	v_{PD}^2-coop	φ_{PD}^2	$v_{Cluster}$	$\varphi_{Cluster}$	$v_{Society}$	$\varphi_{Society}$	Nash
FFF	FFF	FFF	FFF	FFF	MMM	FFF	MMM	N
VHH	VHH	VHH	VHH	VHH	LLL	VHH	HHH	F
HHH	HHH	HHH	HHH	HHH	LLL	HHH	MMM	
MMM	MMM	MMM	MMM	MMM	LLL	HHH	MMM	
LLL	LLL	LLL	LLL	MMM	LLL	HHH	MMM	
VLL	VLL	VLL	VLL	MMM	MMM	MMM	MMM	
NNN	NNN	NNN	NNN	MMM	MMM	MMM	MMM	
$\omega_{cluster}$ = 0.95; $\omega_{Society}$ = 0.05; N_{nn} = 10; $\aleph_c$ = 10								
v_{PD}^1-coop	φ_{PD}^1	v_{PD}^2-coop	φ_{PD}^2	$v_{Cluster}$	$\varphi_{Cluster}$	$v_{Society}$	$\varphi_{Society}$	Nash
FFF	FFF	FFF	FFF	FFF	FFF	FFF	MMM	F-N
VHH	VHH	VHH	VHH	VHH	FFF	VHH	LLL	
HHH	HHH	HHH	HHH	HHH	FFF	HHH	VLL	
MMM	MMM	MMM	MMM	MMM	FFF	MMM	NNN	
LLL	LLL	LLL	LLL	HHH	FFF	MMM	VLL	
VLL	VLL	VLL	VLL	VHH	FFF	LLL	LLL	
NNN	NNN	NNN	NNN	FFF	FFF	MMM	MMM	F
$\omega_{cluster}$ = 0.05; $\omega_{Society}$ = 0.95; N_{nn} = 10; $\aleph_c$ = 10								
v_{PD}^1-coop	φ_{PD}^1	v_{PD}^2-coop	φ_{PD}^2	$v_{Cluster}$	$\varphi_{Cluster}$	$v_{Society}$	$\varphi_{Society}$	Nash
FFF	FFF	FFF	FFF	FFF	MMM	FFF	FFF	F-N
VHH	VHH	VHH	VHH	VHH	LLL	VHH	FFF	F
HHH	HHH	HHH	HHH	HHH	LLL	HHH	FFF	F
MMM	MMM	MMM	MMM	MMM	LLL	HHH	FFF	F
LLL	LLL	LLL	LLL	MMM	LLL	HHH	FFF	F
VLL	VLL	VLL	VLL	MMM	MMM	VHH	FFF	F
NNN	NNN	NNN	NNN	MMM	MMM	FFF	FFF	F-N
$\omega_{cluster}$ = 0.95; $\omega_{Society}$ = 0.95; N_{nn} = 10; $\aleph_c$ = 10								
v_{PD}^1-coop	φ_{PD}^1	v_{PD}^2-coop	φ_{PD}^2	$v_{Cluster}$	$\varphi_{Cluster}$	$v_{Society}$	$\varphi_{Society}$	Nash
FFF	FFF	FFF	FFF	FFF	FFF	FFF	FFF	F-N
VHH	VHH	VHH	VHH	VHH	FFF	VHH	FFF	F
HHH	HHH	HHH	HHH	HHH	FFF	HHH	FFF	F

Table 5 (*continued*)

MMM	MMM	MMM	MMM	MMM	FFF	MMM	FFF	F
LLL	LLL	LLL	LLL	HHH	FFF	HHH	FFF	F
VLL	VLL	VLL	VLL	VHH	FFF	VHH	FFF	F
NNN	NNN	NNN	NNN	FFF	FFF	FFF	FFF	F-N
$\omega_{cluster}$ = 0.5; $\omega_{Society}$ = 0.5; N_{nn} = 10; $\aleph_c$ = 10								
v^1_{PD}-coop	φ^1_{PD}	v^2_{PD}-coop	φ^2_{PD}	$v_{Cluster}$	$\varphi_{Cluster}$	$v_{Society}$	$\varphi_{Society}$	Nash
FFF	FFF	FFF	FFF	VHH	LLL	VHH	HHH	F-N
VHH	VHH	VHH	VHH	HHH	LLL	HHH	HHH	F
HHH	HHH	HHH	HHH	HHH	LLL	HHH	MMM	
MMM	MMM	MMM	MMM	MMM	LLL	HHH	MMM	
LLL	LLL	LLL	LLL	MMM	LLL	HHH	MMM	
VLL	VLL	VLL	VLL	MMM	MMM	MMM	MMM	
NNN	NNN	NNN	NNN	HHH	HHH	HHH	HHH	F

of influence. Second, very large values of orness correspond to situations where the highest linguistic terms contribute more than the others, whereas it is the reverse for lowest values of orness. In other words, low orness favors the lowest values of linguistic feasibility such as NNN and VLL, whereas high orness favors the highest values of linguistic feasibility such as VHH and FFF.

In most cases considered in Tables 5, 6 and 7, there is at least one societal Nash-equilibrium of the F-N, F, or N type. The number of Nash-equilibrium varies with the values of orness within a cluster and at the level of societal game among clusters. It is remarkable that for high values of $\omega_{Society} \gg 0$ we always have seven societal Nash equilibria two of which are of F-N type and the remaining are of F type. For $\omega_{Society} \ll 1$ we obtain only two: one of F type and one of N type. Moreover, for $\omega_{cluster} \ll 1$ we do not get any F-N-Nash equilibrium. We do get one F-N however for $\omega_{cluster} \gg 0$. For medium values of orness $\omega_{Society}$ ~0.5 and $\omega_{cluster}$ ~ 0.5 we obtain two F-Nash equilibria and one F-N-Nash equilibrium.

<u>Internal-PD Game: Two nice-spirited players</u>

In this case (Table 5) I assume that the two players engaged in the PD stage game within the clusters are both nice-spirited; that is, they both perceive full cooperation as fully feasible, very-high cooperation as very-highly feasible, …, no cooperation as not feasible. Yet, they do not necessarily choose the same degrees of nuance and feasibility. For example, one player could opt for a high cooperation with a high degree of feasibility and the other player could choose a low cooperation with low degree of feasibility.

Internal-PD Game: One nice-spirited player against a mean-spirited player

In this case (Table 6) I assume that in the PD stage game within the clusters one player is nice-spirited and the other is mean-spirited. That is, the nice-spirited player initially perceives full cooperation as fully feasible, very-high cooperation as very-highly feasible, …, and null cooperation as not feasible. The mean-spirited player is the mirror image of the nice one; it initially perceives full cooperation as not possible, very-high cooperation with a very low degree of feasibility, …., and null cooperation as highly feasible.

Table 6 Nash social equilibria for cluster-based social game with one nice-spirited player against a mean-spirited player in the internal PD game.

$\omega_{cluster} = 0.05$; $\omega_{Society} = 0.05$; $N_{nn} = 10$; $\aleph_c = 10$								
V^1_{PD}-coop	φ^1_{PD}	V^2_{PD}-coop	φ^2_{PD}	$V_{Cluster}$	$\varphi_{Cluster}$	$V_{Society}$	$\varphi_{Society}$	Nash
FFF	FFF	FFF	NNN	FFF	MMM	FFF	MMM	N
VHH	VHH	VHH	VLL	VHH	LLL	VHH	HHH	F
HHH	HHH	HHH	LLL	HHH	LLL	HHH	MMM	
MMM	MMM	MMM	MMM	MMM	LLL	HHH	MMM	
LLL	LLL	LLL	HHH	HHH	LLL	HHH	MMM	
VLL	VLL	VLL	VHH	VHH	MMM	MMM	MMM	
NNN	NNN	NNN	FFF	FFF	MMM	MMM	MMM	
$\omega_{cluster} = 0.05$; $\omega_{Society} = 0.95$; $N_{nn} = 10$; $\aleph_c = 10$								
V^1_{PD}-coop	φ^1_{PD}	V^2_{PD}-coop	φ^2_{PD}	$V_{Cluster}$	$\varphi_{Cluster}$	$V_{Society}$	$\varphi_{Society}$	Nash
FFF	FFF	FFF	NNN	FFF	MMM	FFF	FFF	F-N
VHH	VHH	VHH	VLL	VHH	LLL	VHH	FFF	F
HHH	HHH	HHH	LLL	HHH	LLL	HHH	FFF	F
MMM	MMM	MMM	MMM	MMM	LLL	HHH	FFF	F
LLL	LLL	LLL	HHH	HHH	LLL	HHH	FFF	F
VLL	VLL	VLL	VHH	VHH	MMM	VHH	FFF	F
NNN	NNN	NNN	FFF	FFF	MMM	FFF	FFF	F-N
$\omega_{cluster} = 0.95$; $\omega_{Society} = 0.05$; $N_{nn} = 10$; $\aleph_c = 10$								
V^1_{PD}-coop	φ^1_{PD}	V^2_{PD}-coop	φ^2_{PD}	$V_{Cluster}$	$\varphi_{Cluster}$	$V_{Society}$	$\varphi_{Society}$	Nash
FFF	FFF	FFF	NNN	FFF	FFF	FFF	MMM	F-N
VHH	VHH	VHH	VLL	VHH	FFF	VHH	LLL	

Table 6 (*continued*)

HHH	HHH	HHH	LLL	HHH	FFF	HHH	VLL	
MMM	MMM	MMM	MMM	MMM	FFF	MMM	NNN	
LLL	LLL	LLL	HHH	HHH	FFF	MMM	VLL	
VLL	VLL	VLL	VHH	VHH	FFF	LLL	LLL	
NNN	NNN	NNN	FFF	FFF	FFF	MMM	MMM	F
$\omega_{cluster} = 0.95$; $\omega_{Society} = 0.95$; $N_{nn} = 10$; $\aleph_c = 10$								
v^1_{PD}-coop	φ^1_{PD}	v^2_{PD}-coop	φ^2_{PD}	$v_{Cluster}$	$\varphi_{Cluster}$	$v_{Society}$	$\varphi_{Society}$	Nash
FFF	FFF	FFF	NNN	FFF	FFF	FFF	FFF	F-N
VHH	VHH	VHH	VLL	VHH	FFF	VHH	FFF	F
HHH	HHH	HHH	LLL	HHH	FFF	HHH	FFF	F
MMM	MMM	MMM	MMM	MMM	FFF	MMM	FFF	F
LLL	LLL	LLL	HHH	HHH	FFF	HHH	FFF	F
VLL	VLL	VLL	VHH	VHH	FFF	VHH	FFF	F
NNN	NNN	NNN	FFF	FFF	FFF	FFF	FFF	F-N
$\omega_{cluster} = 0.5$; $\omega_{Society} = 0.5$; $N_{nn} = 10$; $\aleph_c = 10$								
v^1_{PD}-coop	φ^1_{PD}	v^2_{PD}-coop	φ^2_{PD}	$v_{Cluster}$	$\varphi_{Cluster}$	$v_{Society}$	$\varphi_{Society}$	Nash
FFF	FFF	FFF	NNN	HHH	LLL	VHH	HHH	F-N
VHH	VHH	VHH	VLL	HHH	LLL	HHH	HHH	F
HHH	HHH	HHH	LLL	HHH	LLL	HHH	MMM	
MMM	MMM	MMM	MMM	HHH	LLL	HHH	MMM	
LLL	LLL	LLL	HHH	VHH	LLL	HHH	MMM	
VLL	VLL	VLL	VHH	VHH	MMM	MMM	MMM	
NNN	NNN	NNN	FFF	FFF	HHH	HHH	HHH	F

Internal-PD Game: Two mean-spirited players

In this case (Table 7) I assume that the two players engaged in the PD stage game within the clusters are both mean-spirited; that is, they both initially perceive full cooperation as not feasible, very-high cooperation as very-low feasible, …, null cooperation as feasible.

To sum up: for both dyad-based and cluster-based games we always have at least one F-N Nash equilibrium for intermediate and higher values of orness. Cooperation is always achievable since we always obtain at least a N or F type Nash equilibrium under all situations irrespective of the values of the two types of orness. In short, playing the game of social cooperation under LFL rules predicts the

Table 7 Nash social equilibria for cluster-based social game with two mean-spirited players in the internal PD game.

$\omega_{cluster} = 0.05$; $\omega_{Society} = 0.05$; $N_{nn} = 10$; $\aleph_c = 10$								
v_{PD}^1-coop	φ_{PD}^1	v_{PD}^2-coop	φ_{PD}^2	$v_{Cluster}$	$\varphi_{Cluster}$	$v_{Society}$	$\varphi_{Society}$	Nash
FFF	NNN	FFF	NNN	FFF	MMM	FFF	MMM	N
VHH	VLL	VHH	VLL	VHH	MMM	VHH	MMM	
HHH	LLL	HHH	LLL	HHH	LLL	HHH	MMM	
MMM	MMM	MMM	MMM	MMM	LLL	HHH	MMM	
LLL	HHH	LLL	HHH	MMM	LLL	HHH	MMM	
VLL	VHH	VLL	VHH	LLL	LLL	HHH	HHH	F
NNN	FFF	NNN	FFF	MMM	MMM	MMM	MMM	
$\omega_{cluster} = 0.95$; $\omega_{Society} = 0.05$; $N_{nn} = 10$; $\aleph_c = 10$								
v_{PD}^1-coop	φ_{PD}^1	v_{PD}^2-coop	φ_{PD}^2	$v_{Cluster}$	$\varphi_{Cluster}$	$v_{Society}$	$\varphi_{Society}$	Nash
FFF	NNN	FFF	NNN	FFF	FFF	FFF	MMM	F-N
VHH	VLL	VHH	VLL	VHH	FFF	VHH	LLL	
HHH	LLL	HHH	LLL	HHH	FFF	HHH	VLL	
MMM	MMM	MMM	MMM	MMM	FFF	MMM	NNN	
LLL	HHH	LLL	HHH	HHH	FFF	MMM	VLL	
VLL	VHH	VLL	VHH	VHH	FFF	LLL	LLL	
NNN	FFF	NNN	FFF	FFF	FFF	MMM	MMM	F
$\omega_{cluster} = 0.05$; $\omega_{Society} = 0.95$; $N_{nn} = 10$; $\aleph_c = 10$								
v_{PD}^1-coop	φ_{PD}^1	v_{PD}^2-coop	φ_{PD}^2	$v_{Cluster}$	$\varphi_{Cluster}$	$v_{Society}$	$\varphi_{Society}$	Nash
FFF	NNN	FFF	NNN	FFF	MMM	FFF	FFF	F-N
VHH	VLL	VHH	VLL	VHH	MMM	VHH	FFF	F
HHH	LLL	HHH	LLL	HHH	LLL	HHH	FFF	F
MMM	MMM	MMM	MMM	MMM	LLL	HHH	FFF	F
LLL	HHH	LLL	HHH	MMM	LLL	HHH	FFF	F
VLL	VHH	VLL	VHH	LLL	LLL	VHH	FFF	F
NNN	FFF	NNN	FFF	MMM	MMM	FFF	FFF	F-N
$\omega_{cluster} = 0.95$; $\omega_{Society} = 0.95$; $N_{nn} = 10$; $\aleph_c = 10$								
v_{PD}^1-coop	φ_{PD}^1	v_{PD}^2-coop	φ_{PD}^2	$v_{Cluster}$	$\varphi_{Cluster}$	$v_{Society}$	$\varphi_{Society}$	Nash
FFF	NNN	FFF	NNN	FFF	FFF	FFF	FFF	F-N
VHH	VLL	VHH	VLL	VHH	FFF	VHH	FFF	F
HHH	LLL	HHH	LLL	HHH	FFF	HHH	FFF	F

Table 7 (*continued*)

MMM	MMM	MMM	MMM	MMM	FFF	MMM	FFF	F
LLL	HHH	LLL	HHH	HHH	FFF	HHH	FFF	F
VLL	VHH	VLL	VHH	VHH	FFF	VHH	FFF	F
NNN	FFF	NNN	FFF	FFF	FFF	FFF	FFF	F-N
$\omega_{cluster}$ = 0.5; $\omega_{Society}$ = 0.5; N_{nn} = 10; $\aleph_c$ = 10								
V_{PD}^{1}-coop	φ_{PD}^{1}	V_{PD}^{2}-coop	φ_{PD}^{2}	$V_{Cluster}$	$\varphi_{Cluster}$	$V_{Society}$	$\varphi_{Society}$	Nash
FFF	NNN	FFF	NNN	FFF	HHH	VHH	HHH	F-N
VHH	VLL	VHH	VLL	VHH	MMM	HHH	MMM	
HHH	LLL	HHH	LLL	VHH	LLL	HHH	MMM	
MMM	MMM	MMM	MMM	HHH	LLL	HHH	MMM	
LLL	HHH	LLL	HHH	HHH	LLL	HHH	MMM	
VLL	VHH	VLL	VHH	MMM	LLL	HHH	HHH	F
NNN	FFF	NNN	FFF	MMM	LLL	HHH	HHH	F

emergence of cooperation, with the level of cooperation being moderate or higher and the degree of feasibility being moderate or higher. In certain situations, specifically when $\omega_{Society}$ is large enough, full cooperation is fully feasible along many different pathways (as shown in the various tables).

Chapter 6
Linguistic Fuzzy-Logic and Causality

We depend upon the notion of causation all the time to explain what happens to us. We use causal statements to draw realistic predictions about what might happen as well as to direct what should happen in the future. In short, we are in constant searches for causal explanation and behavior. A widespread motto is to put it formally "X caused Y" or "Y occurred because of X." Not surprisingly then, studying and explaining as well as modeling causality is at the core of much of what counts as social sciences today. Although other types of analysis (such as constitutive) are important aspects of our inquiry about the social and political world, causal analysis absorbs much of our collective effort. Yet, the tools that have been designed for this purpose remain rather limited in social sciences, and more so in political science and international relations. For the most part scholars use statistical means to explore causality. While this has produced a wealth of knowledge and led to the design of powerful statistical tools and insights, there is still a need for other types of tools that would complement the achievements of statistical methods, as well as address issues that statistical analysis cannot reach.

Whether it is statistical/probabilistic or formal/mathematical modeling in nature the analysis of causality assumes a given "logic" and an underlying algebraic structure of this logic. In most studies of causality in social sciences this logic is taken to be the Boolean logic with its associated two-valued algebraic structure. Understanding the inner working of the underlying logic and how it is axiomatically constructed opens up the possibility for considering other types of underlying logics of causal relationships – and there is no shortage of other logics (e.g., fuzzy logic, modal logic, intuitionistic logic, paraconsistent logic, quantum logic, modal logic, …). There are no inherent theoretically, empirically, or methodologically insurmountable impediments or raison d'être why we cannot explore the vast realms of other logics in social science inquiry about causality. In fact, as the burgeoning literature on fuzzy-set theoretic approach to social, economic, and political inquiry testifies, we have much to gain from doing so. It is the goal of this chapter to contribute to this burgeoning literature.

The first section of the chapter presents a Boolean logic approach to causality to pave the way for an LFLA to causality. I then examine causality in social sciences by using propositional calculus in the framework of *linguistic fuzzy logic*. I do this taking into account the possibilities that certain causal variables might be *more or less* of a sufficient type while others might be *more or less* of a necessary type, and while still others might be of both types to a lesser or greater degree of truth. I illustrate the theoretical discussion by examining (following Zinnes'

B. Arfi: Linguistic Fuzzy Logic Methods in Social Sciences, STUDFUZZ 253, pp. 121–154.
springerlink.com

(2004) exploration using binary propositional calculus) the logical validity of the causal arguments made on the theory on the democratic peace. I also apply LFLA analysis to Skocpol's (1979) theory of social revolution (following Goertz and Mahoney, 2005).

1 Analyzing the Logic of Causality

Boolean logic is the most used logic in analyses of causal necessity and sufficiency in social sciences – speaking of logic of causality in social sciences is conventionally tantamount to speaking of Boolean logic of causality. In these analyses of causality, a necessary antecedent X for a consequent Y to occur is expressed as: **Only If X then Y**, whereas a sufficient antecedent is expressed as **If X then Y**. Therefore, the meaning of "implication" as used in formal logic is one of sufficiency. That is, **X→Y is to be understood as: If X then Y** (a presence of X always leads to an occurrence of Y). The necessity clause **Only If X then Y** takes the following formal logic expression: **Y→X**, i.e., any occurrence of Y means that X is also present. Put differently, absent X, Y would not occur, which is expressed formally as **¬X→¬Y**.[1] A clause which states that: a necessary and sufficient antecedent X leads to a consequent Y is expressed as: **If and only if X then Y,** which is expressed in formal logic as: **X↔Y≡(X→Y)∧(¬X→¬Y)** (≡ means the same as). In a summary form we obtain Table 1.

Table 1 Types of Simple Single-Level Causality

Type of Simple Causality	Conventional Semantic	Boolean Logic Form	Alternative Form
Necessity	Only if X then Y	***Y→X***	***¬X→¬Y***
Sufficiency	If X then Y	***X→Y***	***¬Y→¬X***
Necessity and Sufficiency	If and Only if X then Y	***(X→Y)∧(Y→X)***	***(X→Y)∧(¬X→¬Y)***

When one has multiple antecedents (i.e., complex causality) this is conventionally phrased for example as **SW=A*B*C*D** if all antecedents **A, B, C,** and **D** are considered to be necessary conditions. Alternatively, we could have **SW=A*B*C+A*B*D** if, for example, **A, B** and **C** are jointly necessary conditions and **A, B,** and **D** are jointly necessary conditions (and a number of other possible combinations). The conjoint necessary condition clause **SW=A*B*C*D** actually means **Only If (A and B and C and D) then SW**, i.e., **SW→A∧B∧C∧D,** or equivalently using the law of contraposition **¬(A∧B∧C∧D)→¬SW**. Likewise, the conjointly sufficient condition clause **SW=A*B*C+A*B*D** would really mean **If ((A and B and C) or (A and B and D)) then SW**, i.e., **(A∧B∧C)∨(A∧B∧D)→SW**.

How would causal inferences at two different levels determine inference between two levels? Simple two-level causality occurs, for example, through a

[1] This comes from the law of contraposition of Boolean logic: **(A→B) ↔ (¬B→¬A).**

propagation of **logical inference** between two necessary inferences as in the following example:

$$\{\text{Only If X then Y}\}\ \text{Implies}\ \{\text{Only If Y then Z}\} \qquad (108)$$

That is, X is a necessary antecedent for Y, Y is a necessary antecedent for Z, and the fact that X is a necessary antecedent for Y sufficiently implies that Y is a necessary for Z. Symbolically:

$$(\neg X \rightarrow \neg Y) \rightarrow (\neg Y \rightarrow \neg Z) \qquad (109)$$

This is one example of a spectrum of possibilities of two-level inferences. The variation occurs because we can have logical necessity or sufficiency within both levels of causality as well as necessity or sufficiency at the level of the inference itself linking the two levels. Moreover, the situation can become even more complicated by assuming causal complexity (a combination of necessary and sufficient antecedents) at both levels. Thus, it will be simple minded to assume for example that the logic of two-level causality can be exhaustively studied by considering the following form only:

$$X \rightarrow Y \rightarrow Z \qquad (110)$$

This is only one case out a set of 27 cases (see Table 2). More specifically, this latter form can be interpreted as:

$$(X \rightarrow Y) \rightarrow (Y \rightarrow Z) \qquad (111)$$

That is:

$$\text{If (Only If X then Y) then (Only If Y then Z)} \qquad (112)$$

This assumes that we have all logical inferences – at the first level, second level, and the across-level inference – of a sufficiency type. Another pitfall in the analysis of the logic of causality can occur if one defines the meaning of two-level causality (X causes Y which causes Z) as the same as:

$$(X \rightarrow Y) \wedge (Y \rightarrow Z) \qquad (113)$$

This is not wrong since in Boolean logic, which is generally taken as the underlying logic in most social sciences inquiries, **(X→Y)Λ(Y→Z)** is equivalent through the law of syllogism to **X→Z.** The latter is the commonsensical meaning of causality in social sciences. It might however lead to a circumvented investigation of the logic of two-level causality. Indeed, defining two-level causality as **(X→Y)Λ(Y→Z)** does make it easy to explore the type of causality (i.e., in terms of necessity and sufficiency), which is an essential task in most causal analysis in social sciences. However, this is in no way an exhaustive study of all possibilities.

Therefore, when logically examining simple two-level causality we must consider three sorts or levels of logical inference: within the first and second levels of inference and the across-level inference. For example, necessary inference at the three inference levels reads as:

$$\{\neg\,(\neg X \rightarrow \neg Y)\} \rightarrow \{\neg\,(\neg Y \rightarrow \neg Z)\} \tag{114}$$

That is:

$$\text{Only If (Only If X then Y) then (Only If Y then Z).} \tag{115}$$

Likewise, necessity and sufficiency at all levels of inference reads as:

$$\left[(X \rightarrow Y) \wedge (\neg X \rightarrow \neg Y)\right] \wedge \left\{\neg\left[(Y \rightarrow Z) \wedge (\neg Y \rightarrow \neg Z)\right]\right\} \tag{116}$$

That is:

$$\text{If and Only If (If and Only If X then Y) then (If and Only If Y then Z)} \tag{117}$$

Note that Boolean logic possesses a number of laws that can be used to simplify these expressions of two-level causality. This is not without a cost, though. Whereas the final expressions might be formally simpler, the connection with empirically-relevant semantic meaning in terms of necessity and sufficiency might not be readily extracted from the final simplified logical formula. Therefore, for the sake of empirically useful semantic interpretation, which is an important step in social science analysis, it might at times be more beneficial to work with the non-simplified versions of logical formulas.

In sum, Boolean analysis of causality has a number of powerful features. First, it clearly explicates the logic of combining different independent variables. Second, Boolean analysis clearly and consistently explicates necessary and sufficient conditions as well as various combinations thereof. Third, the process of Boolean minimization helps to systematically isolate relevant from irrelevant independent variables. Fourth, it also includes counterfactual analysis through the negation operator which strengthens the validity of the causal analysis.

Linguistic fuzzy-logic analysis of causality consists of asking a number of IF-Then fuzzy questions to reach a conclusion about the combination of the different dimensions of causality. For example, we would have the following fuzzy inference rule:

$$\text{IF}\{A \Diamond B \Diamond C \Diamond D\}\text{THEN SW} \tag{118}$$

I use the $\Diamond$ symbol instead of AND or OR because in linguistic fuzzy logic complex causality is neither strictly about necessity (which is usually represented with a logical AND and symbolically by a product of the independent variables) nor strictly about sufficiency (which is usually represented with a logical OR and

Table 2 Types of Simple Two-Level Causality (N=Necessary; S=Sufficient; NS=N & S)

1st Level Inference Logic Formula		2nd Level Inference Logic Formula		1st → 2nd Level Inference Logic Formula		
N	Only If X then Y ¬X→¬Y	N	Only If Y then Z ¬Y→¬Z	N	Only If (Only If X then Y) then (Only If Y then Z) ¬ (¬X→¬Y) → ¬ (¬Y→¬Z)	1
N	Only If X then Y ¬X→¬Y	N	Only If Y then Z ¬Y→¬Z	S	If (Only If X then Y) then (Only If Y then Z) (¬X→¬Y) → (¬Y→¬Z)	2
N	Only If X then Y ¬X→¬Y	N	Only If Y then Z ¬Y→¬Z	NS	If and Only If (Only If X then Y) then (Only If Y then Z) {(¬X→¬Y) → (¬Y→¬Z)}∧{¬(¬X→¬Y) → ¬(¬Y→¬Z)}	3
N	Only If X then Y ¬X→¬Y	S	If Y then Z Y→Z	N	Only If (Only If X then Y) then (If Y then Z) ¬ (¬X→¬Y) → ¬ (Y→Z)	4
N	Only If X then Y ¬X→¬Y	S	If Y then Z Y→Z	S	If (Only If X then Y) then (If Y then Z) (¬X→¬Y) → (Y→Z)	5
N	Only If X then Y ¬X→¬Y	S	If Y then Z Y→Z	NS	If and Only If (Only If X then Y) then (If Y then Z) {(¬X→¬Y)→(Y→Z)}∧{¬ (¬X→¬Y)→¬ (Y→Z)}	6
N	Only If X then Y ¬X→¬Y	NS	If and Only If Y then Z (Y→Z)∧(¬Y→¬Z)	N	Only If (Only If X then Y) then (If and Only If Y then Z) ¬(¬X→¬Y) →¬((Y→Z)∧(¬Y→¬Z))	7
N	Only If X then Y ¬X→¬Y	NS	If and Only If Y then Z (Y→Z)∧(¬Y→¬Z)	S	If (Only If X then Y) then (If and Only If Y then Z) (¬X→¬Y) →((Y→Z)∧(¬Y→¬Z))	8
N	Only If X then Y ¬X→¬Y	NS	If and Only If Y then Z (Y→Z)∧(¬Y→¬Z)	NS	If and Only If (Only If X then Y) then (If and Only If Y then Z) {(¬X→¬Y)→((Y→Z)∧(¬Y→¬Z))} ∧ {¬(¬X→¬Y)→¬((Y→Z)∧(¬Y→¬Z))}	9
S	If X then Y X→Y	N	Only If Y then Z ¬Y→¬Z	N	Only If (If X then Y) then (Only If Y then Z) ¬ (X→Y) → ¬ (¬Y→¬Z)	10
S	If X then Y X→Y	N	Only If Y then Z ¬Y→¬Z	S	If (If X then Y) then (Only If Y then Z) (X→Y) → (¬Y→¬Z)	11
S	If X then Y X→Y	N	Only If Y then Z ¬Y→¬Z	NS	If and Only If (If X then Y) then (Only If Y then Z) {(X→Y)→(¬Y→¬Z)} ∧ {¬(X→Y)→ ¬(¬Y→¬Z)}	12
S	If X then Y X→Y	S	If Y then Z Y→Z	N	Only If (If X then Y) then (If Y then Z) ¬ (X→Y) → ¬ (Y→Z)	13
S	If X then Y X→Y	S	If Y then Z Y→Z	S	If (If X then Y) then (If Y then Z) (X→Y) → (Y→Z)	14
S	If X then Y X→Y	S	If Y then Z Y→Z	NS	If and Only If (If X then Y) then (If Y then Z) {(X→Y)→(Y→Z)} ∧ {¬(X→Y)→ ¬(Y→Z)}	15
S	If X then Y X→Y	NS	If and Only If Y then Z (Y→Z)∧(¬Y→¬Z)	N	Only If (If X then Y) then (If and Only If Y then Z) ¬(X→Y) → ¬((Y→Z)∧(¬Y→¬Z))	16
S	If X then Y X→Y	NS	If and Only If Y then Z (Y→Z)∧(¬Y→¬Z)	S	If (If X then Y) then (If and Only If Y then Z) (X→Y) → ((Y→Z)∧(¬Y→¬Z))	17
S	If X then Y X→Y	NS	If and Only If Y then Z (Y→Z)∧(¬Y→¬Z)	NS	If and Only If (If X then Y) then (If and Only If Y then Z) {(X→Y) → (Y→Z)∧(¬Y→¬Z)} ∧ { ¬(X→Y) → ¬((Y→Z)∧(¬Y→¬Z))}	18

Table 2 (*continued*)

NS	If and Only If X then Y $(X\to Y)\wedge(\neg X\to\neg Y)$	N	Only If Y then Z $\neg Y\to\neg Z$	N	Only If (If and Only If X then Y) then (Only If Y then Z) $\neg((X\to Y)\wedge(\neg X\to\neg Y))\to\neg(\neg Y\to\neg Z)$	19
NS	If and Only If X then Y $(X\to Y)\wedge(\neg X\to\neg Y)$	N	Only If Y then Z $\neg Y\to\neg Z$	S	If (If and Only If X then Y) then (Only If Y then Z) $((X\to Y)\wedge(\neg X\to\neg Y))\to(\neg Y\to\neg Z)$	20
NS	If and Only If X then Y $(X\to Y)\wedge(\neg X\to\neg Y)$	N	Only If Y then Z $\neg Y\to\neg Z$	NS	If and Only If (If and Only If X then Y) then (Only If Y then Z) $\{((X\to Y)\wedge(\neg X\to\neg Y))\to(\neg Y\to\neg Z)\}\wedge\{\neg((X\to Y)\wedge(\neg X\to\neg Y))\to\neg(\neg Y\to\neg Z)\}$	21
NS	If and Only If X then Y $(X\to Y)\wedge(\neg X\to\neg Y)$	S	If Y then Z $Y\to Z$	N	Only If (If and Only If X then Y) then (If Y then Z) $\neg((X\to Y)\wedge(\neg X\to\neg Y))\to\neg(Y\to Z)$	22
NS	If and Only If X then Y $(X\to Y)\wedge(\neg X\to\neg Y)$	S	If Y then Z $Y\to Z$	S	If (If and Only If X then Y) then (If Y then Z) $((X\to Y)\wedge(\neg X\to\neg Y))\to(Y\to Z)$	23
NS	If and Only If X then Y $(X\to Y)\wedge(\neg X\to\neg Y)$	S	If Y then Z $Y\to Z$	NS	If and Only If (If and Only If X then Y) then (If Y then Z) $\{((X\to Y)\wedge(\neg X\to\neg Y))\to(Y\to Z)\}\wedge\{\neg((X\to Y)\wedge(\neg X\to\neg Y))\to\neg(Y\to Z)\}$	24
NS	If and Only If X then Y $(X\to Y)\wedge(\neg X\to\neg Y)$	NS	If and Only If Y then Z $(Y\to Z)\wedge(\neg Y\to\neg Z)$	N	Only If (If and Only If X then Y) then (If and Only If Y then Z) $\neg((X\to Y)\wedge(\neg X\to\neg Y))\to\neg((Y\to Z)\wedge(\neg Y\to\neg Z))$	25
NS	If and Only If X then Y $(X\to Y)\wedge(\neg X\to\neg Y)$	NS	If and Only If Y then Z $(Y\to Z)\wedge(\neg Y\to\neg Z)$	S	If (If and Only If X then Y) then (If and Only If Y then Z) $((X\to Y)\wedge(\neg X\to\neg Y))\to((Y\to Z)\wedge(\neg Y\to\neg Z))$	26
NS	If and Only If X then Y $(X\to Y)\wedge(\neg X\to\neg Y)$	NS	If and Only If Y then Z $(Y\to Z)\wedge(\neg Y\to\neg Z)$	NS	If and Only If (If and Only If X then Y) then (If and Only If Y then Z) $\{((X\to Y)\wedge(\neg X\to\neg Y))\to((Y\to Z)\wedge(\neg Y\to\neg Z))\}\wedge\{\neg((X\to Y)\wedge(\neg X\to\neg Y))\to\neg((Y\to Z)\wedge(\neg Y\to\neg Z))\}$	27

symbolically by an addition of the independent variables).[2] Linguistic fuzzy logic allows a formalization of fuzzy causality as "more or less necessary like" and "more or less sufficient like," usually termed as approximate reasoning in fuzzy logic literature.

A key question raised by linguistic fuzzy logic analysis of simple two-level causality is the extent to which it is empirically relevant to have a variety of Boolean logic forms of two-level causality (27 cases in total, see Table 2). Many of these forms have to do with the fact that we assume a dichotomy between necessary and sufficient antecedents. As discussed shortly, linguistic fuzzy logic analysis of causality partially avoids such a dichotomy through the use of generalized notions of logical connectives such as linguistic disjunction and conjunction. In applying this to simple two-level causality we need to keep in mind that we have three sorts of inferences: first-level inference, second-level inference, and across-level inference. One immediate implication of this is that not only the logical connectives AND and OR are generalized in the same level but also across levels. This means that just as keeping a dichotomy between necessity and sufficiency within one level of causality is rather untenable (in fuzzy logic) it is also untenable in across-level causality. In other words, we cannot crisply differentiate between causality operating at one level and then a second one at another level – the two

[2] See for example: Goertz (2003:68-70).

inextricably more or less contaminate one another. We should hence speak of fuzzy causality *tout court* – not two-level causality.

To sum up at this point, we need to recast the formal expressions for simple causality (for both necessity and sufficiency) in such a way as to take into account three essential features of a linguistic fuzzy logic approach. First, all antecedent and consequent variables are linguistically expressed and symbolically manipulated. Second, the truth values of these antecedent and consequent variables as well as all combinations thereof formed by using the four connectives ($\wedge, \vee, \rightarrow, \neg$) are also linguistically expressed and symbolically manipulated. Third, these truth values are many-valued. This calls both for the firm axiomatic basis developed in Chapter 2 for manipulating these linguistic truth values and for a procedure for rigorously manipulating all symbolic variables and truth values.

As a starting point for the analysis let's assume that we have for example a double logical inference A $\rightarrow$ B $\rightarrow$ C. Let us formulate this using the language of linguistic fuzzy logic:

1. *We know that: {If "x is A" then "y is B"} is true to a degree* τ_β
2. *We know that: {If "y is B" then "z is C"} is true to a degree* τ_γ
3. *We know that: {"x is A" is true to a degree* τ_α*}*
4. *To what degree of truth* τ_δ *can we deduce that:*

   If [{If "x is A" then "y is B"} is true to a degree τ_β *]*

   Then [{If "y is B" then "z is C"} is true to a degree τ_γ *]?*

Put differently: We know the linguistic value of A and its truth value τ_α. We know the truth values of both inference processes 1 and 2, respectively, τ_β and τ_γ. Can we infer the truth value τ_δ of the implication **A causes B which then causes C**? In solving this problem we need to apply the procedure for one level to three levels of inference. That is: we need to do two tasks at each level of inference: a first one on how to aggregate the linguistic information on the relevance levels of the antecedents and a second one on how to aggregate the linguistic information on the truth values of the antecedents. When studying the type of causality (that is, whether it is of a necessary, sufficient or combined type) we need to keep in mind that the antecedents at the second level causality are not simply the consequents at the first level of causality. As discussed earlier, we can have 27 combinations of necessary and/or sufficiency combinations because we are in fact combining three sets of cardinality 3 each, that is, {N,S,NS} within first level, {N,S,NS} within second level, and {N,S,NS} across levels. The linguistic fuzzy logic problem posed just before the beginning of this paragraph is an instance of [S,S,S]. An instance of [N,N,N] would assume the following form:

1. *We know that: {Only If "x is A" then "y is B"} is true to a degree* τ_β
2. *We know that: {Only If "y is B" then "z is C"} is true to a degree* τ_γ

3. *We know that: {"x is A" is true to a degree* τ_α*}*
4. *To what degree of truth* τ_δ *can we deduce that*

   Only If [{Only If "x is A" then "y is B"} is true to a degree τ_β *]*

   Then [{Only If "y is B" then "z is C"} is true to a degree τ_γ *]?*

Likewise, for [NS,NS,NS] we get:

1. *We know that: {If and Only If "x is A" then "y is B"} is true to a degree* τ_β
2. *We know that: {If an Only If "y is B" then "z is C"} is true to a degree* τ_γ
3. *We know that: {"x is A" is true to a degree* τ_α*}*
4. *To what degree of truth* τ_δ *can we deduce that*

   If and Only If [{If and Only If "x is A" then "y is B"} is true to a degree τ_β *]*

   Then [{If and Only If "y is B" then "z is C"} is true to a degree τ_γ *]?*

In translating these inference clauses into formal language of linguistic fuzzy logic we need the three connectives: (linguistic conjunction) $\wedge$, (linguistic implication) $\rightarrow$, and (linguistic negation) $\neg$. As discussed earlier, we can study linguistic conjunction and disjunction using the operators LWC, LWD, and LWA. Linguistic negation is formalized by using Neg(.) as defined earlier. The linguistic implication connective is formalized by using LWI (linguistic weighted implication). The formalism is developed next through an illustrative example, i.e., testing the logical consistency of the so-called democratic peace argument of international relations theory.

2 Linguistic Fuzzy-Logic Analysis of the Democratic Peace Argument

In this section I consider Zinnes' (2004) study of the logical consistency of the democratic peace argument. Using Boolean propositional logic, Zinnes formalizes the propositions of the democratic peace theory by first constructing the premises (propositions) for the normative argument as shown in Table 3. A democratic state is denoted by A1 and a non-democratic state is denoted by A2. A dyad of two democratic states is denoted by:

$$\mathrm{A1} \wedge \neg \mathrm{A2} \tag{119}$$

A dyad with one democratic state and one non-democratic state is denoted by:

$$\mathrm{A1} \wedge \mathrm{A2} \text{ or alternatively by } \neg\, \mathrm{A1} \wedge \neg \mathrm{A2} \tag{120}$$

Table 3 Zinnes' Basic Propositions for Democratic Peace

Basic Propositions for Normative Explanation	Notation
One democratic state	A1
One non-democratic state	A2
All state decisions involve the participation of the population and its representative institutions	A3
All decisions are made by a small group of elite leaders	A4
Conflict over societal policies is resolved through bargaining	A5
Conflict over societal politics is resolved using force	A6
State uses force to settle internation conflicts	A7
State uses a bargaining strategy to settle internation conflicts	A8
State's security is threatened	A9
States are in conflict	A10
Both states use force to settle their conflict	A11
States go to war	A12

A dyad with two non-democratic states is denoted by:

$$\neg A1 \wedge A2 \tag{121}$$

Zinnes also postulates a number of intermediary premises shown in Table 4.

Table 4 Zinnes' Postulated Premises for Democratic Peace

Postulated Premises for Normative Explanation	Notation
If state X is a democracy, then all decisions involve the participation of the population and its representative institutions	N1: $A1 \rightarrow A3$
If state Y is a democracy, then all decisions involve the participation of the population and its representative institutions	N2: $\neg A2 \rightarrow A3$
If state Y is a non-democracy, then all decisions are made by a small group of elite leaders	N3: $A2 \rightarrow A4$
If state X is a non-democracy, then all decisions are made by a small group of elite leaders	N4: $\neg A1 \rightarrow A4$
If all decisions involve the participation of the population and its representative institutions, then conflicts over societal policies are resolved through bargaining	N5: $A3 \rightarrow A5$
If all decisions are made by a small group of elite leaders, then conflicts over societal policies are resolved using force	N6: $A4 \rightarrow A6$
If a state uses force to settle conflicts over societal policies, then it uses force to settle internation conflicts	N7: $A6 \rightarrow A7$
If a state uses bargaining to settle internal societal conflicts and its security is not threatened, then it will use bargaining to settle internation conflicts	N8: $A5 \wedge \neg A9 \rightarrow A8$
If a state's security is threatened, then it will use force to settle internation conflicts	N11: $A9 \rightarrow A7$
If states X and Y are in conflict and one state's security is threatened, then states X and Y use force to settle the conflict	N12: $A10 \wedge A9 \rightarrow A11$
If states X and Y use force to settle internation conflict, then states X and Y go to war	N13: $A11 \rightarrow A12$

Table 4 (*continued*)

If states X and Y are in conflict and state X is a democracy and state Y is democracy, then force is not used to settle internation conflicts	**N14: A10 ∧ A1 ∧ ¬A2 → ¬A7**
If states X and Y are in conflict and a state's security is not threatened, then force is not used to settle internation conflicts	**N15: A10 ∧ ¬A9 → ¬A7**
If states X and Y are in conflict and bargaining is used to settle internation conflicts, then states X and Y do not use force to settle internation conflicts	**N16: A10 ∧ A8 → ¬A11**
If states X and Y do not use force to settle internation conflict, then states X and Y do not go to war	**N17: ¬A11→ ¬A12**

Doing the Boolean propositional calculus and postulating **A10** (two states in conflict) leads to the conclusions on the normative explanation as shown in Table 5.

Table 5 Zinnes' Two-Valued Logical Pathways for Democratic Peace Argument

Two democracies (A1 ∧ ¬A2) ∧ A10	Democracy and non-democracy: (A1 ∧ A2) ∧ A10;	(¬A1 ∧ ¬A2) ∧ A10	Two non-democracies: (¬A1 ∧ A2) ∧ A10
N1: A1 → A3 N2: ¬A2 → A3	N1: A1 → A3 N3: A2 → A4	N2: ¬A2 → A3 N4: ¬A1 → A4	N3: A2 → A4 N4: ¬A1 → A4
N5: A3 → A5	N6: A4 → A6		
N8: A5 ∧ ¬A9 → A8	N7: A6 → A7		
N11: A9 → A7	N10: A10 ∧ A7 → A9		
N14: A10 ∧ A1 ∧ ¬A2 → ¬A7	N12: A10 ∧ A9 → A11		
N16: A10 ∧ A8 → ¬A11	N13: A11 → A12 (War)		
N17: ¬A11→ ¬A12 (No War)			

Let us now generalize Zinnes' analysis. First, we assume that a state is not just either democratic or non-democratic. Rather, we postulate that being democratic (or non-democratic) is a linguistic variable the value of which belongs to the set of linguistic labels:[3]

$$S = \{s_i \mid i = 0,1,2,3,4\}, \text{ with}$$

$$s_0 = NNN, s_1 = LMM, s_2 = MMM, s_3 = MHH, s_4 = FFF$$

Hence, a state can have a null (NNN) degree of being democratic (completely non-democratic), a low to moderate degree of being democratic (LMM), a moderate degree of being democratic (MMM), a moderate to high degree of being democratic (MHH), or a full degree (FFF) of being democratic (fully democratic). We also assume that each of these levels or degrees of being democratic has a truth value, which is also one of the elements of S. For example, we could have a state with a MMM degree of being democratic and the truth value of this knowledge is LMM. Likewise, we fuzzify all of the basic propositions used by Zinnes in her exploration of the democratic peace. By "fuzzify" I mean that every

[3] I am assuming this small set for the sake of making the illustration easy to follow.

Table 6 Fuzzy Propositions for Democratic Peace Argument

Basic Fuzzy Propositions	Notation
One more or less democratic state	LF1
All state decisions more or less involve the participation of the population and its representative institutions	LF3
Conflict over societal policies is more or less resolved through bargaining	LF5
State more or less uses force to settle internation conflicts	LF7
State's security is more or less threatened	LF9
States are more or less in conflict	LF10
Both states more or less use force to settle their conflict	LF11
States more or less go to war	LF12

Table 7 Postulated Linguistic Fuzzy Premises for Democratic Peace

Postulated Linguistic Fuzzy Premises	Denotation of Linguistic Fuzzy Proposition:
X is more or less democratic and Y is more or less democratic and X and Y are more or less in conflict, with each feeling more or less threatened	LFP1[X] ∧ LFP1[Y] ∧ LFP10[X,Y] ∧ LFP9[Y] ∧ LFP9[Y]
If state is a more or less democracy, then all decisions more or less involve the participation of the population and its representative institutions	LFP1: LF1 → LF3
If all decisions more or less involve the participation of the population and its representative institutions, then conflicts over societal policies are more or less resolved through bargaining	LFP5: LF3 → LF5
If a state more or less uses force to settle conflicts over societal policies, then it more or less uses force to settle internation conflicts	LFP7: LF5 → F7
If a state more or less uses bargaining to settle internal societal conflicts and its security is more or less threatened, then it will more or less use bargaining to settle internation conflicts	LFP8: LF5 ∧ LF9 → LF7
If a state's security is more or less threatened, then it will more or less use force to settle internation conflicts	LFP11: LF9 → LF7
If states X and Y are more or less in conflict and one state's security is more or less threatened, then states X and Y more or less use force to settle the conflict	LFP12: LF10 ∧ LF9 → LF11
If states X and Y more or less use force to settle internation conflict, then states X and Y more or less go to war	LFP13: LF11 → LF12
If states X and Y are more or less in conflict and state X is a more or less democracy and state Y is more or less democracy, then force is more or less used to settle internation conflicts	LFP14: LF10 ∧ LF1[X] ∧ LF1[Y] → LF7

Table 7 (*continued*)

If states X and Y are more or less in conflict and a state's security is more or less not threatened, then force is more or less used to settle internation conflicts	**LFP15: LF10 ∧ LF9 → LF7**
If states X and Y are more or less in conflict and bargaining is more or less used to settle internation conflicts, then states X and Y more or less use force to settle internation conflicts	**LFP16: LF10 ∧ LF7 → LF11**
If states X and Y more or less use force to settle internation conflict, then states X and Y more or less go to war	**LFP17: LF11→ LF12**

proposition has a linguistic value and that each of these linguistic values has a linguistic truth value attached to it. We hence obtain Tables 6 and 7.

Table 7 assumes a number of input "variables"; these are:

1. *LF1[X]: degree of democracy of state X*
2. *LF1[Y]: degree of democracy of state Y*
3. *LF10[X,Y]: degree of conflict between states X and Y*
4. *LF9[X]: degree of threat felt by state X*
5. *LF9[Y]: degree of threat felt by state Y*

Checking the logic of the democratic peace argument using a linguistic fuzzy approach is based on positing linguistic values for the truth values of these variables as well as the truth values of the connectives used in the propositions. The question is thus:

- Given LFP1[X], LFP1[Y], LFP10[X,Y], LFP9[X] and LFP9[Y],
- Given the linguistic fuzzy propositions LFP1, …, LFP17,
- What is the truth value of the argument that "more or less democratic states more or less do not fight war?"

In the end what we really are exploring is the following inference:

$$(\text{LF1[X]} \wedge \text{LF1[Y]} \wedge \text{LF10[X,Y]} \wedge \text{LFP9[Y]} \wedge \text{LFP9[Y]}) \rightarrow \text{LFP12[X,Y]?} \quad (122)$$

We use the LWI and LWC operators where the linguistic weighted implication function LWI stands for the following linguistic fuzzy inference:

- *Given that A and B are linguistic terms*
- *We know that: {"X is A" is true to a degree* τ_α*}*
- *To what degree of truth* τ_δ *can we infer that: LWI (X,Y):*

If {"X is A" with a truth value τ_α*} then {"Y is B" with a truth value* τ_β*}?*

The linguistic weighted conjunction LWC is given by:

- *Given that A and B are linguistic terms*
- *We know that: {"X is A" is true to a degree* τ_α*}*
- *We know that: {"Y is B" is true to a degree* τ_β*}*
- *What is the degree of truth* τ_δ *of LWC(X,Y): Linguistic Conjunction of X and Y?*

The fuzzy inference process consists of five phases (I, II, III, IV, V) as shown in Figure 1 and explicated in Figure 2. The procedure is to start with a set of given premises and then evaluate the truth value of the whole process starting from phase I to phase V. That is:

- *Given: {A=* $LF1[X], LF10[X,Y], LF9[X]$*} and {B=* $LF1[Y], LF10[X,Y], LF9[Y]$ *}*
- *What is the truth value of {(A and B)* $\rightarrow$ *more or less WAR}?*

A number of points are in order. First, these implications and conjunctions are linguistic-fuzzy in the sense that their truth values are linguistic values expressed using a chosen linguistic-terms set. Second, we have (as shown in Figure 1) a number of intervening implications and conjunctions the truth values of which are also linguistic terms. This means that the truth value of the whole logical argument is determined by the initial premises as well as by the intervening logical operations. Third, it is not meaningful to divide the possible initial sets of dyads into three categories as Zinnes does in her Boolean-based analysis. That is: instead of (Democracy, Democracy), (non-Democracy, Democracy) and (non-Democracy, non-Democracy) we have many more possibilities of (more or less democracy, more or less democracy). For example in the set:

$$S = \{s_i \mid i = 0,1,2,3,4\} \tag{123}$$

$$s_0 = NNN, s_1 = LMM, s_2 = MMM, s_3 = MHH, s_4 = FFF \tag{124}$$

We would have 5x5=25 dyads, some of which are: (NNN,NNN), (LMM,MMM), (FFF, MHH), with NNN democracy = no democracy, LMM democracy = low to moderate level of democracy, MHH democracy = moderate to high level of democracy, and FFF democracy = full democracy.

Fourth, because we are dealing with linguistic truth values, the democratic peace argument can be asserted to have a full degree of validity (FFF), a moderate to high degree of validity (MHH), a moderate degree of validity (MMM), a low to moderate degree of validity (LMM), or a null degree of validity (NNN). This degree of validity is obtained at the phase V of the process of linguistic fuzzy analysis (shown in Figures 1 and 2). In Table 8, I display a number of illustrations obtained through a MATLab computer code. The democratic peace argument expressed in a linguistic formalism states that:

$$INFERENCE \quad (IF\{\text{LF1}[X] = FFF \text{ and LF1}[Y] = FFF\} \quad THEN \quad \{\text{LF12}[X,Y] = NNN\}) \quad is \quad \text{TRUE}$$
$$INFERENCE \quad (IF\{\text{LF1}[X] = FFF \text{ and LF1}[Y] = NNN\} \quad THEN \quad \{\text{LF12}[X,Y] = FFF\}) \quad is \quad \text{TRUE} \tag{125}$$

Written in linguistic fuzzy formalism, the Boolean logic premises and conclusions are listed for in Table 9.

Table 8 Linguistic Fuzzy Testing of Democratic Peace Argument (DPA)

Linguistic Values of Various Propositional Premises										Truth Value of DPA
Row	LF1[X]	LF1[Y]	LF10[X,Y]	LF9[X]	LF3[X]	LF5[X]	LF7[X]	LF11[X,Y]	LF12[X,Y]	LWI
1	FFF	FFF	MMM	LMM	FFF	FFF	FFF	LMM	LMM	MHH
2	FFF	FFF	MHH	MMM	FFF	FFF	FFF	MMM	LMM	MMM
3	FFF	FFF	NNN	NNN	FFF	FFF	FFF	NNN	NNN	FFF
4	FFF	NNN	MHH	MHH	FFF	FFF	FFF	MMM	MMM	MMM
5	FFF	NNN	LMM	MMM	FFF	FFF	MMM	LMM	MMM	MHH
6	MMM	MMM	MHH	LMM	MMM	LMM	LMM	MHH	MHH	MHH
7	MMM	MMM	LMM	LMM	MMM	MMM	MMM	MHH	FFF	FFF
8	LMM	MHH	FFF	MHH	MMM	LMM	LMM	MHH	LMM	MHH
9	NNN	NNN	MHH	MHH	NNN	NNN	LMM	MHH	FFF	FFF
10	NNN	NNN	NNN	NNN	NNN	NNN	MMM	MHH	FFF	FFF
11	NNN	NNN	FFF	FFF	NNN	NNN	NNN	FFF	FFF	FFF

Table 9 Boolean Testing of Democratic Peace Argument (DPA)

Linguistic Values of Various Propositional Premises				DPA
Row	STATE [X]	STATE [Y]	WAR [X,Y]	Truth Value
1	FFF = DEMOCRATIC	FFF = DEMOCRATIC	FFF = WAR	NNN = FALSE
2	FFF = DEMOCRATIC	FFF = DEMOCRATIC	NNN = NO WAR	FFF = TRUE
3	FFF = DEMOCRATIC	NNN = NON-DEMOCRATIC	FFF = WAR	FFF = TRUE
4	FFF = DEMOCRATIC	NNN = NON-DEMOCRATIC	NNN = NO WAR	FFF = FALSE
5	NNN = NON-DEMOCRATIC	NNN = NON-DEMOCRATIC	FFF = WAR	FFF = TRUE
6	NNN = NON-DEMOCRATIC	NNN = NON-DEMOCRATIC	NNN = NO WAR	FFF = FALSE

We see that the linguistic fuzzy analysis reproduces the Boolean result – Row 3 in Table 8 and Row 2 in Table 9, respectively – that two democratic states do not fight one another. We also see that both analyses – Row 11 in Table 8 and Row 1

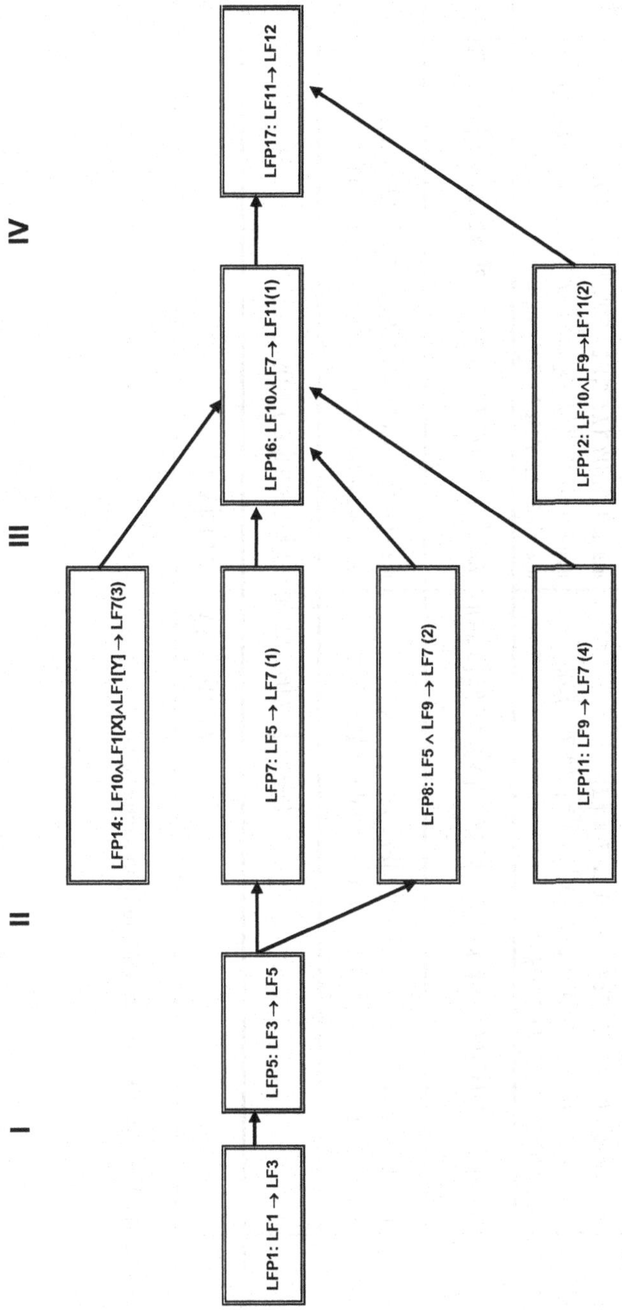

Fig. 1 Fuzzy Causal Inferences in Democratic Peace

	State X:	State Y:
	Given: $LF1[X], LF10[X,Y], LF9[X]$	Given: $LF1[Y], LF10[X,Y], LF9[Y]$
I	I: $LWI\{LF1[X], LF3[X]\}$	I: $LWI\{LF1[Y], LF3[Y]\}$
II	II: $LWI\{LF3[X], LF5[X]\}$	II: $LWI\{LF3[Y], LF5[Y]\}$
III	III.1: $LWI\{LF5[X], LF7[X](1)\}$ III.2: $LWI\{LWC(LF5[X], LF9[X]), LF7[X](2)\}$ III.3: $LWI\{LWC(LF10[X,Y], LF1[X], LF1[Y]), LF7[X](3)\}$ III.4: $LWI\{LF9[X], LF7[X](4)\}$	III.1: $LWI\{LF5[Y], LF7[Y](1)\}$ III.2: $LWI\{LWC(LF5[Y], LF9[Y]), LF7[Y](2)\}$ III.3: $LWI\{LWC(LF10[X,Y], LF1[X], LF1[Y]), LF7[Y](3)\}$ III.4: $LWI\{LF9[Y], LF7[Y](4)\}$
	$LF7[X] = LWC\{LF7[X](1), LF7[X](2), LF7[X](3), LF7[X](4)\}$	$LF7[Y] = LWC\{LF7[Y](1), LF7[Y](2), LF7[Y](3), LF7[Y](4)\}$
IV	IV.1: $LWI\{LWC(LF10[X,Y], LF7[X]), LF11[X,Y](1)\}$ IV.2: $LWI\{LWC(LF10[X,Y], LF9[X]), LF11[X,Y](2)\}$	IV.1: $LWI\{LWC(LF10[X,Y], LF7[Y]), LF11[X,Y](1)\}$ IV.2: $LWI\{LWC(LF10[X,Y], LF9[Y]), LF11[X,Y](2)\}$
	$LF11[X,Y] = LWC\{LF11[X,Y](1), LF11[X,Y](2)\}$	
V	V: $LWI\{LF11[X,Y], LF12[X,Y]\}$	

Fig. 2 The Process of Fuzzy Inference in Democratic Peace

in Table 9 – agree that two non-democratic states fight one another.[4] Some clarification is needed concerning the various columns of Table 8. Take the case of Row 3 which agrees with its counterpart from Boolean analysis. The meanings of the various symbols in Row 3 are as follows:

- *LF1[X] = FFF means that state X is fully democratic*
- *LF1[Y] means that state Y is fully democratic*
- *LF10[X,Y] = NNN means the truth value that there is a conflict between X and Y is null*
- *LF9[X] = NNN means the truth value that state X feels threatened is null*
- *LF12[X,Y] = FFF means the truth value that there is war between X and Y is null, i.e., No War.*

The remaining variables, i.e., LF3[X], LF5[X], LF7[X], and LF11[X,Y] are intermediate inferences in the logical argument. Each is assumed to be true in the Boolean analysis (not shown in Table 9). In the linguistic fuzzy analysis the truth values of these inferences can vary. In Row 3 of Table 8 they all equal FFF, i.e., they are all true. Changing the truth values of these intermediate inferences while keeping those for the first two variables (i.e., still considering a dyad of two fully democratic states), changes the truth value of the democratic peace argument. In other words: the argument is not simply either true (or false) but becomes partially true, as shown in Rows 1 and 2 of Table 8. For example (Row 2), for a dyad of two fully democratic states who believe with a moderate-to-high degree of truth that there is a conflict between them, and among which X (or Y or both) believes with a low to moderate degree of truth that its security is threatened, the democratic peace argument is moderately true. Row 7 of Table 8 is another interesting one. For a dyad of moderately democratic states, the analysis predicts that War (LF12[X,Y]=FFF) can occur with a full degree of truth (FFF)! In sum, while the linguistic fuzzy analysis does not prove the democratic peace argument to be wrong, *it adds much nuance to it.* We do not have only options of War and Non-War. We also have many more possibilities of more or less war and no-war depending on whether the states are more or less democratic. This linguistic fuzzy logic conclusion seems to provide a logical explanation (although not from the conventional perspective of Boolean logic) to the still ongoing debate on the empirical validity of the democratic peace argument.

One can arguably say that: the problem seems neither to be one of empirical validation, nor one of theoretical explanation (although these are still somewhat debatable). Instead, from the perspective of this chapter, the problem seems to have much to do with the taken for granted Boolean logic approach which underpins the existing literature on the democratic peace!

---

[4] Note that in the linguistic fuzzy approach all four ways of computing the implication function agree in each case (that is, all LWIs produce FFF in both types of dyads), except for LWI4 in Row 11 of Table 8.

3 Linguistic Fuzzy-Logic Testing of Skocpol Theory of Revolution

This section uses the LFLA to check the causal consistency of typical social science theory, namely, Skocpol's theory of revolution. As argued by Goertz and Mahoney (2005), Skocpol's (1979) theory of revolution is based on a two-level causal argument. In this section I apply the linguistic fuzzy-logic approach to this two-level analysis of causality. The starting point of the two-level analysis is that we have a double logical inference such as X → Y → Z. Let us phrase this using the language of linguistic fuzzy logic:

1. We know that:
   The inference statement:
   *S1= {**If** "X has a nuance value* V_x *to a level of truth* τ_x*" then "Y has a nuance value* V_y *to a level of truth* τ_y*"} has a level of nuanced inference* V_α *and a degree of truth* τ_α
2. We know that:
   The inference statement:
   *S2= {**If** "Y has a nuance value* V_y *to a level of truth* τ_y*" then "Z has a nuance value* V_z *to a level of truth* τ_z*"} has a level of nuanced inference* V_β *and a degree of truth* τ_β
3. To what level of nuanced inference V_γ and what degree of truth τ_γ can we infer that:
   ***If** [statement S1 has a level of nuanced inference* V_α *and a degree of truth* τ_α*] then [statement S2 has a level of nuanced inference* V_β *and a degree of truth* τ_β*]?*

Put differently: We know the nuance and truth values of both inference processes 1 and 2, respectively, (V_α, τ_α)and (V_β, τ_β), can we infer the nuance and truth values (V_γ, τ_γ) of the implication **{X causes Y which then causes Z}**?

When studying two-level causality we need to keep in mind that we can have 27 combinations of necessity and/or sufficiency relations because we are in fact combining three sets each with cardinality 3, that is, {N,S,NS} within the first level, {N,S,NS} within the second level, and {N,S,NS} across levels, where N=necessity, S=sufficiency, NS=necessity and sufficiency. The inference (using linguistic fuzzy logic) stated just before the beginning of this paragraph is an instance of [S,S,S]. An instance of [N,N,N] would assume the following form:

1. We know that:
*The inference statement S1= {**Only If** "X has a nuance value* V_x *to a level of truth* τ_x*" then "Y has a nuance value* V_y *to a level of truth* τ_y*"} has a level of nuanced inference* V_α *and a degree of truth* τ_α
2. We know that:
*The inference statement S2= {**Only If** "Y has a nuance value* V_y *to a level of truth* τ_y*" then "Z has a nuance value* V_z *to a level of truth* τ_z*"} has a level of nuanced inference* V_β *and a degree of truth* τ_β
3. To what level of nuanced inference V_γ and what degree of truth τ_γ can we infer that:
***Only If** [statement S1 has a level of nuanced inference* V_α *and a degree of truth* τ_α*] then [statement S2 has a level of nuanced inference* V_β *and a degree of truth* τ_β*]?*

Likewise, for [NS,NS,NS] we get:

1. We know that:
*The inference statement S1= {**If and Only If** "X has a nuance value* V_x *to a level of truth* τ_x*" then "Y has a nuance value* V_y *to a level of truth* τ_y*"} has a level of nuanced inference* V_α *and a degree of truth* τ_α
2. We know that:
*The inference statement S2= {**If and Only If** "Y has a nuance value* V_y *to a level of truth* τ_y*" then "Z has a nuance value* V_z *to a level of truth* τ_z*"} has a level of nuanced inference* V_β *and a degree of truth* τ_β
3. To what level of nuanced inference V_γ and what degree of truth τ_γ can we infer that:
***If and Only If** [statement S1 has a level of nuanced inference* V_α *and a degree of truth* τ_α*] then [statement S2 has a level of nuanced inference* V_β *and a degree of truth* τ_β*]?*

As discussed earlier, we study linguistic conjunction, disjunction, and implication using the operators: LWC, LWD, LWA, and LI.

In the remainder of the section I apply the LFLA procedure of analyzing multi-path causation to Skocpol's theory of social revolution. I closely follow the causal paths explicated by Goertz and Mahoney (2005) and displayed in Figure 3.

Skocpol defines social revolutions as "rapid, basic transformations of a society's state and class structures; and they are accompanied by and in part carried through by class-based revolts from below" (1979:4-5). Therefore, social revolutions (**SRV**) are the combination of three attributes: (1) **CRV** = class-based revolt from below; (2) **TSS** = rapid and basic transformation of state structures; (3) **TCS** = rapid and basic transformation of class structures. These attributes are **individually necessary and jointly sufficient** for social revolution to occur (Goertz and Mahoney, 2005). This relationship is not one of causality but rather one of ontology. I denote this part of the argument as ONTA.

Skocpol's theory has two levels of causality. At the basic level (denoted from here and on as BCAUS), the argument consists of a conjuncture of two necessary causes which are jointly sufficient for social revolutions to occur. These are: **PRV** = *peasant revolt* and **SBR** = *state breakdown*. As put by Skocpol (1979: 154), "I have argued that (1) state organizations susceptible to administrative and military collapse when subjected to intensified pressures from more developed countries from abroad, and (2) agrarian sociopolitical structures that facilitated widespread peasant revolts against landlords were, taken together, the sufficient distinctive causes of social-revolutionary situations commencing in France, 1789, Russia, 1917, and China, 1911." Hence, as stated by Goertz and Mahoney (2005), "state breakdown and peasant revolt are *individually necessary and jointly sufficient* for social revolution" (emphasis added).

At the secondary level of causality (denoted as SCAUS), Skocpol considers the different processes that might lead to sate breakdown and peasant revolt. At this stage of the argument, Skocpol is actually looking at an across-level causal relationship between the secondary-level variables and the basic-level ones effectively arguing that the *second-level variables are individually sufficient but not necessary to cause either state breakdown or peasant revolt.* In explicating the first-level causes of state breakdown, Skocpol considers three different second-level causes: (1) **INT** = *international pressure*, which promotes crises for the actors of the regime; (2) **DCL** = *dominant-class leverage* within the state, which prevents government leaders from implementing modernizing reforms; and (3) **AGB** = *agrarian backwardness*, which hinders national state responses to political crises. To explicate the causes of peasant revolt, Skocpol considers two secondary level variables: (1) **PAS** = *peasant autonomy and solidarity*, which facilitate spontaneous collective action by peasants; and (2) **LAV** = *landlord vulnerability*, which allows for class transformation in the countryside.

Skocpol's argument is that at the basic level, PRV and SBR are individually necessary and jointly sufficient to cause CSR (social revolution), GM express this as:

$$CSR = PRV * SBR \tag{126}$$

To express this causal relationship using formal logic we need to keep in mind that the statement "*individually necessary and jointly sufficient*" is effectively a conjunction of three statements:

S1: SBR is a necessary cause of CSR: $CSR \rightarrow SBR$ *or, the logical equivalent,* $\neg SBR \rightarrow \neg CSR$

S2: PRV is a necessary cause of CSR: $CSR \rightarrow PRV$ *or, the logical equivalent,* $\neg PRV \rightarrow \neg CSR$

S3: PRV and SBR jointly sufficiently cause CSR: $(SBR \wedge PRV) \rightarrow CSR$

Table 10 Skocpol's Theory Variables

Variable	Meaning
CSR	Causal Relationship - Social Revolution
SBR	State Breakdown
INT	International Pressure
DCL	Dominant-Class Leverage
AGB	Agrarian Backwardness
PRV	Peasant Revolt
PAS	Peasant Autonomy
LAV	Landlord Vulnerability
OSR	Ontological Relationship - Social Revolution
CRV	Class Revolts
TSS	Transformation of State Structure
TCS	Transformation of Class Structure

Therefore, the complete formal logic expression for [126] is:

$$S1 \wedge S2 \wedge S3 = \{CSR \rightarrow SBR\} \wedge \{CSR \rightarrow PRV\} \wedge \{(PRV \wedge SBR) \rightarrow CSR\} \tag{127}$$

The membership function for the logical implication $P \rightarrow Q$ is given by:

$$\mu_{P \rightarrow Q} = \max\left[1 - \mu_P ; \mu_Q\right] \tag{128}$$

μ_P is the membership function for proposition P, and $1 - \mu_P$ is the corresponding one for the negation of P, non-P. Using this we obtain the following expressions for the membership functions:

$$\mu_{CSR \rightarrow SBR} = \max\left[1 - \mu_{CSR} ; \mu_{SBR}\right] \tag{129}$$

$$\mu_{CSR \rightarrow PRV} = \max\left[1 - \mu_{CSR} ; \mu_{PRV}\right] \tag{130}$$

$$\mu_{SBR \wedge PRV \rightarrow CSR} = \max\left[1 - \mu_{SBR \wedge PRV} ; \mu_{CSR}\right] \tag{131}$$

$$\text{With } \mu_{SBR \wedge PRV} = \min\left[\left(\mu_{SBR} + \mu_{PRV}\right) ; 1\right] \tag{132}$$

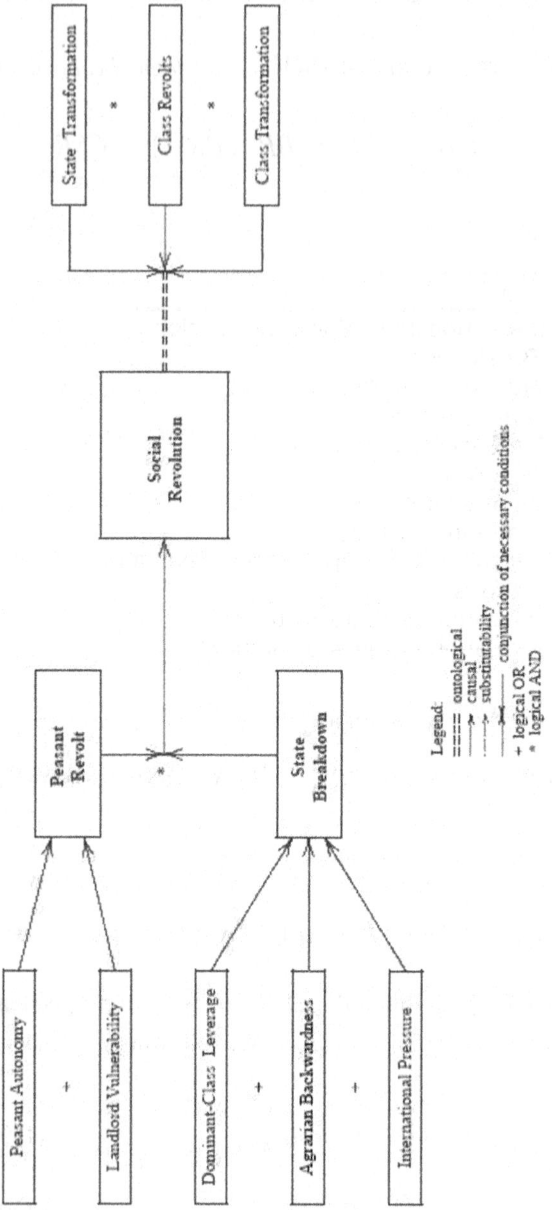

Fig. 3 Skocpol's Causal Theory of Social Revolutions[5]

[5] Goertz and Mahoney (2005:509).

The membership degree of the conjunction of the three statements – $S1 \wedge S2 \wedge S3$ – is thus obtained as:

$$\mu_{S1\wedge S2\wedge S3} = \min\left(\left(\mu_{S1} + \mu_{S2} + \mu_{S3}\right);1\right) = \min\left(\left(\mu_{CSR\to SBR} + \mu_{CSR\to PRV} + \mu_{SBR\wedge PRV\to CSR}\right);1\right) \tag{133}$$

Combining together equations[128], [129] – [133], we obtain the logically correct membership function for [127] – that is, at the basic level of causality BCAUS as:

$$\begin{aligned} \mu^{BCAUS} = \;& \min\left\{\left(\max\left[1-\mu_{CSR};\mu_{SBR}\right] + \max\left[1-\mu_{CSR};\mu_{PRV}\right]\right.\right. \\ & \left.\left. + \max\left[1-\min\left(\left(\mu_{SBR}+\mu_{PRV}\right);1\right);\mu_{CSR}\right]\right);1\right\} \end{aligned} \tag{134}$$

Skocpol's argument at the secondary level is that INT, AGB, and DCL are jointly sufficient. GM's shorthand for this would be:

$$SBR{=}INT{+}AGB{+}DCL \tag{135}$$

In the formalism of formal logic we would have:

$$(INT \to SBR) \vee (AGB \to SBR) \vee (DCL \to SBR) \tag{136}$$

In the language of fuzzy-set theory (that GM use) equation [136] would correspond to a membership degree:

$$\begin{aligned} \mu^{SCAUS}(SBR) &= \max\left\{\mu_{INT\to SBR};\mu_{AGB\to SBR};\mu_{DCL\to SBR}\right\} \\ &= \max\left\{\max\left[1-\mu_{INT};\mu_{SBR}\right];\max\left[1-\mu_{AGB};\mu_{SBR}\right];\max\left[1-\mu_{DCL};\mu_{SBR}\right]\right\} \end{aligned} \tag{137}$$

Likewise, PAS and LAV are jointly sufficient, thus GM's shorthand expression for this is:

$$PRV{=}PAS + LAV \tag{138}$$

In formal logic we get:

$$(PAS \to PRV) \vee (LAV \to PRV) \tag{139}$$

In the language of fuzzy-set theory (that GM use) equation [139] would correspond to a membership degree:

$$\begin{aligned} \mu^{SCAUS}(PRV) &= \max\left\{\mu_{PAS\to PRV};\mu_{LAV\to PRV}\right\} \\ &= \max\left\{\max\left[1-\mu_{PAS};\mu_{PRV}\right];\max\left[1-\mu_{LAV};\mu_{PRV}\right]\right\} \end{aligned} \tag{140}$$

At the ontological level GM argue that TSS, TCS, and CRV are *individually necessary and jointly sufficient* for social revolution to occur, which in their shorthand notation takes the form:

$$OSR = TSS * TCS * CRV \tag{141}$$

Expressed in formal logic, the ontological relationship thus assumes the form:

$$(OSR \rightarrow TSS) \wedge (OSR \rightarrow TCS) \wedge (OSR \rightarrow CRV) \tag{142}$$

The corresponding membership function is:

$$\mu^{ONTA} = \min\{\max[1-\mu_{OSR};\mu_{TSS}] + \max[1-\mu_{OSR};\mu_{TCS}] + \max[1-\mu_{OSR};\mu_{CRV}];1\} \tag{143}$$

From equations [134], [137], [140], and [143] we see that the membership functions μ_{CSR}, μ_{SBR}, μ_{PRV}, μ_{OSR} for CRS at the basic level of causality, SBR and PRV at the secondary level of causality, and OSR at the ontological level are *inputs that need to be estimated from the empirical data.*

Having explicated the membership functions corresponding to the various levels of analysis that GM delineate in Skocpol's theory, it should be clear that the logical basis of GM's coding and testing of Skocpol's theory is flawed. For example, when testing the logic of causality for CSR in terms of SBR and PRV, GM compute the membership function μ_{CSR} of the variable CSR in terms of the membership functions μ_{SRB} and μ_{PRV} of SBR and PRV as:

$$\mu_{CSR} = \mu_{SBR \wedge PRV} = \mu_{SBR} \cdot \mu_{PRV} \tag{144}$$

GM then use this as measure to test causality in Skocpol's theory at the basic level. This formula is erroneously based on an incorrect logical argument of the causality statement. The problem is that GM take the shorthand equation **CSR=SBR*PRV** for a formal logic equation. This is not true even within the framework of Boolean logic as Table 11 shows. We can clearly see that the logical values of the two formal logic expressions

$$\text{SBR} \wedge \text{PRV} \rightarrow \text{CSR} \tag{145}$$

$$(\text{CSR} \rightarrow \text{SBR}) \wedge (\text{CSR} \rightarrow \text{PRV}) \wedge [(\text{SBR} \wedge \text{PRV}) \rightarrow \text{CSR}] \tag{146}$$

do not always agree with one another (see row 5 in Table 11 where the discrepancy occurs).

What gets lost in using the shorthand equation **CSR=SBR*PRV** is the role that logical implication plays in the causal relationship. The logically correct equation

Table 11 Boolean-Logic Truth Table for Basic Causality Level Leading to CSR (T=true, F=false).

Row	CSR	SBR	PRV	CSR→ SBR	CSR → PRV	SBR ∧ PRV	SBR∧PRV → CSR	(CSR → SBR) ∧(CSR → PRV)	(CSR → SBR) ∧(CSR → PRV) ∧ (SBR∧PRV → CSR)
1	T	T	T	T	T	T	T	T	T
2	T	T	F	T	F	F	T	F	F
3	T	F	T	F	F	F	T	F	F
4	T	F	F	F	F	F	T	F	F
5	F	T	T	T	T	T	F	T	F
6	F	T	F	T	T	F	F	T	F
7	F	F	T	T	T	F	F	T	F
8	F	F	F	T	T	F	F	T	F

Table 12 GM's Fuzzy Coding of Skocpol's Data at Secondary Level (expressed in GM's numerical and LFLA formalisms)

	INT		DCL		AGB		PAS		LAV	
France	0.50	MMM	0.75	HHH	1.00	PPP	0.75	HHH	1.00	PPP
Russia 1917	1.00	PPP	0.25	LLL	0.50	MMM	1.00	PPP	1.00	PPP
China	0.75	HHH	0.75	HHH	1.00	PPP	0.00	NNN	0.75	HHH
England	0.50	MMM	1.00	PPP	0.25	LLL	0.00	NNN	0.00	NNN
Russia 1905	0.50	MMM	0.25	LLL	0.50	MMM	1.00	PPP	1.00	PPP
Germany	0.25	LLL	0.25	LLL	0.25	LLL	0.50	MMM	0.00	NNN
Prussia	0.75	HHH	0.25	LLL	0.25	LLL	0.50	MMM	0.00	NNN
Japan	0.75	HHH	0.00	NNN	0.50	MMM	0.00	NNN	0.00	NNN

Table 13 GM's Fuzzy Coding of Skocpol's Theory at the Basic Causal Level CSR (expressed in GM numerical and LFLA formalisms)

	SBR		PRV		CSR			
					Minimum		Min(Sum Xi,1)	
France	1.00	PPP	1.00	PPP	1.00	PPP	1.00	PPP
Russia 1917	1.00	PPP	1.00	PPP	1.00	PPP	1.00	PPP
China	1.00	PPP	0.75	HHH	1.00	PPP	1.00	PPP
England	1.00	PPP	0.00	NNN	0.00	NNN	0.42	MLL
Russia 1905	0.50	MMM	1.00	PPP	0.00	NNN	0.33	MLL
Germany	0.25	LLL	0.50	MMM	0.00	NNN	0.17	VLL
Prussia	0.75	HHH	0.50	MMM	0.00	NNN	0.25	LLL
Japan	0.75	HHH	0.00	NNN	0.00	NNN	0.42	MLL

can be obtained by going back to the logical meaning of "logical implication: →", the formal logic formula for which is: $\{P \rightarrow Q\}$ or, equivalently, $\{\neg P \vee Q\}$, which reads as: P implies Q, or, equivalently, non-P (negation of P) OR Q with the membership function given by equation [128]. Moreover, GM's method seeks to

Table 14 GM's Fuzzy Coding of Skocpol's Theory at the Ontological Level (expressed in GM numerical and LFLA formalisms)

	CRV		TSS		TCS		OSR Minimum		OSR Min(Sum Xi,1)	
France	1.00	PPP	1.00	PPP	1.00	PPP	1.00	PPP	1.00	PPP
Russia 1917	1.00	PPP	1.00	PPP	1.00	PPP	1.00	PPP	1.00	PPP
China	1.00	PPP	1.00	PPP	1.00	PPP	1.00	PPP	1.00	PPP
England	0.00	NNN	1.00	PPP	0.25	LLL	0.00	NNN	0.42	MLL
Russia 1905	1.00	PPP	0.00	NNN	0.00	NNN	0.00	NNN	0.33	MLL
Germany	0.50	MMM	0.00	NNN	0.00	NNN	0.00	NNN	0.17	VLL
Prussia	0.00	NNN	0.25	LLL	0.50	MMM	0.00	NNN	0.25	LLL
Japan	0.00	NNN	1.00	PPP	0.25	LLL	0.00	PPP	0.42	MLL

evaluate the membership functions for the variables CSR, SRB, PRV, and OSR as a way for testing Skocpol's theory. This is incorrect from a formal logic perspective since as shown up above the membership functions μ_{CSR}, μ_{SBR}, μ_{PRV}, μ_{OSR} are inputs, not outputs. This is in line with Ragin's original procedure as developed in his 2000 book. As will discussed in Chapter 7, Ragin's method for testing causality takes as inputs the membership functions of all independent and dependent variables.

In sum, to test causality we need to infer the membership functions of formal logic expressions that are faithful translations of Skocpol's arguments (as excellently delineated by GM in natural language) at the basic causality, secondary causality, and ontological levels. What GM need to do (following Ragin's fuzzy-set theoretic approach) is to empirically estimate the values for μ_{CSR}, μ_{SBR}, μ_{PRV}, μ_{OSR} and then use equations [134], [137], [140], and [143] to estimate the various arguments. Tables 12, 13, and 14 respectively show GM's estimated input data at the secondary causality, basic causality, and ontological levels. Note that GM estimate the membership functions for CSR and OSR (that is, the outcome variables at basic causality and ontological levels) in two ways: (1) dichotomously in line for Skocpol's own estimate – France, Russia 1917 and China had successful revolutions while the other cases did not); (2) using the family resemblance rule formalized through the equation $\min\left(Sum\ X_i;1\right)$. I also use both in the LFLA method to test Skocpol's theory.

In view of these remarks, I thus compute what a corrected GM approach would predict for the various processes – basic and secondary causality as well as ontological relationship. These results are then compared to what LFLA predicts. The results of the corrected GM – denoted from here and onward as CGM – fuzzy testing of Skocpol's theory are shown in Table 15.

One unexpected result is that I obtain exactly the same results for the four processes whether I use GM's coding for OSR and CSR with the *min* or *min(SumXi,1)* rule. According to Table 15, there is a perfect (PPP) agreement between what the corrected GM fuzzy-testing predicts using GM's input data and Skocpol's arguments on the outcome variables OSR, PRV, and CSR for all states. In the case of

Table 15 Corrected GM's Fuzzy Testing of Skocpol's Theory (expressed in GM numerical and LFLA formalisms)

Process	Ontological		Secondary Causality				Basic Causality			
Outcome Variable	ONTA		SCSBR		SCPRV		BCAUS			
							CGM		GM: SBR*PRV	
France	1	PPP	1	PPP	1	PPP	1	PPP	1.00	PPP
Russia 1917	1	PPP	1	PPP	1	PPP	1	PPP	1.00	PPP
China	1	PPP	1	PPP	1	PPP	1	PPP	0.75	HHH
England	1	PPP	1	PPP	1	PPP	1	PPP	0.00	NNN
Russia 1905	1	PPP	0.75	HHH	1	PPP	1	PPP	0.50	MMM
Germany	1	PPP	0.75	HHH	1	PPP	1	PPP	0.25	LLL
Prussia	1	PPP	0.75	HHH	1	PPP	1	PPP	0.50	MMM
Japan	1	PPP	1	PPP	1	PPP	1	PPP	0.00	NNN

the outcome variable SBR, the corrected GM approach predicts less than a perfect process of secondary causality BCSBR with a level of HHH (high) causality. The largest disparity between the GM result and CGM approach occurs in the prediction at the level of basic causality BCAUS with the outcome variable CSR (see Table 15). However, making such a comparison does not make much sense – it is like comparing oranges and potatoes. As pointed out earlier, GM err at the level of formal logic and hence their results are not a direct test of basic causality *per se*. What they predict is the value of the outcome variable CSR by using the formula CSR=SBR*PRV. The corrected GM method does not predict the values of the outcome variables – rather, the latter are used as input data. The CGM method will instead be compared to the LFLA method, to which I turn next.

In order to use LFLA in a logically consistent way and hence avoid the error made by GM, we need to start from the formal logic equations [127], [136], [139], and [142] in combination with the connectives of linguistic fuzzy logic, LWC, LWD, LI and LWA. At the basic causality process BCAUS leading to CSR, equation [127] takes the form

$$\left(v_{BCAUS},\tau_{BCAUS}\right)=LWC\left\{\left(v_{CSR\to SBR},\tau_{CSR\to SBR}\right);\left(v_{CSR\to PRV},\tau_{CSR\to PRV}\right);\left(v_{SBR\wedge PRV\to CSR},\tau_{SBR\wedge PRV\to CSR}\right)\right\} \tag{147}$$

Where:

$$\left(v_{CSR\to SBR},\tau_{CSR\to SBR}\right)=LI\left\{\left(v_{CSR},\tau_{CSR}\right);\left(v_{SBR},\tau_{SBR}\right)\right\} \tag{148}$$

$$\left(v_{CSR\to PRV},\tau_{CSR\to PRV}\right)=LI\left\{\left(v_{CSR},\tau_{CSR}\right);\left(v_{PRV},\tau_{PRV}\right)\right\} \tag{149}$$

$$\begin{aligned}\left(v_{SBR\wedge PRV\to CSR},\tau_{SBR\wedge PRV\to CSR}\right)&=LI\left\{\left(v_{SBR\wedge PRV},\tau_{SBR\wedge PRV}\right);\left(v_{CSR},\tau_{CSR}\right)\right\}\\&=LI\left\{LWC\left\{\left(v_{SBR},\tau_{SBR}\right);\left(v_{PRV},\tau_{PRV}\right)\right\};\left(v_{CSR},\tau_{CSR}\right)\right\}\end{aligned} \tag{150}$$

At the secondary causality process SCSBR leading to SBR, equation [136] takes the form:

$$(\nu_{BCSBR},\tau_{BCSBR}) = LWD\{(\nu_{INT\to SBR},\tau_{INT\to SBR});(\nu_{AGB\to SBR},\tau_{AGB\to SBR});(\nu_{DCL\to SBR},\tau_{DCL\to SBR})\} \quad (151)$$

Where, using X for INT, AGB, or DCL, we have:

$$(\nu_{X\to SBR},\tau_{X\to SBR}) = LI\{(\nu_X,\tau_X);(\nu_{SBR},\tau_{SBR})\} \quad (152)$$

At the secondary causality process SCPRV leading to PRV, equation [139] takes the form:

$$(\nu_{BCPRV},\tau_{BCPRV}) = LWD\{(\nu_{PAS\to PRV},\tau_{PAS\to PRV});(\nu_{LAV\to PRV},\tau_{LAV\to PRV})\} \quad (153)$$

Where, using X for PAS or LAV, we have:

$$(\nu_{X\to PRV},\tau_{X\to PRV}) = LI\{(\nu_X,\tau_X);(\nu_{PRV},\tau_{PRV})\} \quad (154)$$

Finally, at the ontological level ONTA leading to the outcome variable OSR, equation [142] leads to:

$$(\nu_{ONTA},\tau_{ONTA}) = LWC\{(\nu_{OSR\to TSS},\tau_{OSR\to TSS});(\nu_{OSR\to TCS},\tau_{OSR\to TCS});(\nu_{OSR\to CRV},\tau_{OSR\to CRV})\} \quad (155)$$

Where, using X for TSS, TCS, or CRV, we have:

$$(\nu_{OSR\to X},\tau_{OSR\to X}) = LI\{(\nu_{OSR},\tau_{OSR});(\nu_X,\tau_X)\} \quad (156)$$

Equations [147], [150], [151], [153], and [155] can be generalized by using the aggregation operation LWA instead of LWC and LWD for various values of orness ω.[6]

As explained in the previous chapters, LFLA requires inputs and produces outputs for nuance levels ν_i and truth values τ_i. We thus face the same situation as in the previous chapters – we have an added degree of freedom that we can use to directly probe the causal process. However, this calls for specific input data on truth values, which GM do not provide in their rendering of Skocpol's theory and data. I hence first assume that the degrees of truth of all variables are maximum, that is, PPP. I then consider two additional cases: one with uniformly randomized truth values (given in Table 24) and one with MHH truth values for GM's input data to highlight the role that truth values play – should play – in analyzing empirical data using LFLA. As explained earlier in the book, ν_i for a process would be an estimate of fuzzy process (causality or ontological relation) whereas

[6] See Chapter 2 for a discussion of orness.

τ_i would be an estimate of our confidence on this estimate. The results of the analysis are displayed in Tables 16 – 23.

Table 16 compares the results obtained using CGM (corrected GM analysis) and LFLA at the secondary causal process SCSBR leading to SBR for two different values of orness (ω~0.3, 0.8), while assuming perfect (PPP) truth degrees (or levels of confidence) for the input variables that GM deduce from Skocpol's empirical study of the eight states. The table presents the obtained values for the nuance and truth degrees in the case of LFLA using the LWD (linguistic weighted disjunction) and LWA (linguistic weighted average which is a generalization of LWD). There is a perfect agreement between CGM and LFLA at low levels of orness for France, Russia in 1917, China and England. That is: there is perfect causality PPP at this level. These results are predicted with a perfect (PPP) level of confidence for a low level of orness. At high levels of orness LFLA thru LWA still agrees with CGM but LFLA thru LWD does not. France and China show a lower degree of nuanced causality (VHH) than Russia-1917 and England (which still have a PPP degree). For high level of orness we only have a VHH (very high) confidence level in the prediction. A noticeable result is that all four cases of LFLA predict a MMM (moderate) nuanced causality with an MMM level of confidence for Russia-1905, a value which is much less than the value MMM that CGM predicts. Likewise, all LFLA cases agree with CGM prediction of high (HHH) nuanced causality with a HHH level of confidence for Prussia. In the case of Germany, all LFLA cases predict a low (LLL) nuanced causality whereas CGM predicts a high (HHH) nuanced causality. In the case of Japan, all LFLA cases predict a high (HHH) nuanced causality with HHH level of confidence, whereas CGM predicts a PPP nuanced causality. We can conclude from this analysis that Skocpol's theory does a very good job in explaining the secondary causality level leading to SBR in the cases of France, Russia-1917, China and England. However, the posited secondary causality leading to SBR does not seem to fare well in the cases of Russia 1905, Germany, Prussia, and Japan. Yet it is important to note that the conclusions are not dichotomous. Even in the cases where there were no social revolutions, the secondary level causality is more or less operational, depending on specificities of the cases.

Table 16 Comparing CGM and LFLA Analyses of Skocpol's Theory: Secondary Causal Process SCSBR Leading to SBR for ω~0.3, 0.8 and PPP input truth values.

				SCSBR								
	INT	DCL	AGB	CGM	LWD: ω~0.3		LWD: ω~0.8		LWA: ω~0.3		LWA: ω~0.8	
State	ν	ν	ν	ν	ν	τ	ν	τ	ν	τ	ν	τ
France	MMM	HHH	PPP	PPP	PPP	PPP	VHH	VHH	PPP	PPP	PPP	VHH
Russia 1917	PPP	LLL	MMM	PPP	PPP	PPP	PPP	VHH	PPP	PPP	PPP	VHH
China	HHH	HHH	PPP	PPP	PPP	PPP	VHH	VHH	PPP	PPP	PPP	VHH
England	MMM	PPP	LLL	PPP	PPP	PPP	PPP	VHH	PPP	PPP	PPP	VHH
Russia 1905	MMM	LLL	MMM	HHH	MMM	MMM	MMM	MMM	MMM	MMM	MMM	MMM
Germany	LLL	LLL	LLL	HHH	LLL	HHH	LLL	MHH	LLL	LLL	LLL	LLL
Prussia	HHH	LLL	LLL	HHH	HHH	HHH	HHH	HHH	HHH	HHH	HHH	HHH
Japan	HHH	NNN	MMM	PPP	HHH	HHH	HHH	HHH	HHH	HHH	HHH	HHH

Table 17 compares the results obtained using CGM (corrected GM analysis) and LFLA at the secondary causal process SCPRV leading to PRV for two different values of orness (ω~0.3, 0.8), while assuming perfect (PPP) truth degrees (or levels of confidence). One clearly noticeable feature of these results is that CGM predicts for all states a PPP nuanced secondary causality leading to PRV. LFLA for low value of orness through LWD and for both low and high values of orness through LWA agrees with CGM. At a low value of orness LFLA through LWD predicts a high (HHH) nuanced causality only. As to the levels of confidence of these LFLA predictions, we have a perfect (PPP) level of confidence at low level of orness and a high (HHH) level of confidence for high level of orness only. The conclusion is that at this level of secondary causality, the process depicted by Skocpol's theory seems to be operational in all states at a level of at least HHH nuanced causality, with at least a HHH level of confidence, at both low and high levels of orness.

Table 17 Comparing CGM and LFLA Analyses of Skocpol's Theory: Secondary Causal Process SCPRV Leading to PRV for ω~0.3, 0.8 and PPP input truth values.

			SCPRV								
	PAS	LAV	CGM	LWD: ω~0.3		LWD: ω~0.8		LWA: ω~0.3		LWA: ω~0.8	
State	ν	ν	ν	ν	τ	ν	τ	ν	τ	ν	τ
France	HHH	PPP	PPP	PPP	PPP	HHH	HHH	PPP	PPP	PPP	HHH
Russia 1917	PPP	PPP	PPP	PPP	PPP	HHH	HHH	PPP	PPP	PPP	HHH
China	NNN	HHH	PPP	PPP	PPP	HHH	HHH	PPP	PPP	PPP	HHH
England	NNN	NNN	PPP	PPP	PPP	HHH	HHH	PPP	PPP	PPP	HHH
Russia 1905	PPP	PPP	PPP	PPP	PPP	HHH	HHH	PPP	PPP	PPP	HHH
Germany	MMM	NNN	PPP	PPP	PPP	HHH	HHH	PPP	PPP	PPP	HHH
Prussia	MMM	NNN	PPP	PPP	PPP	HHH	HHH	PPP	PPP	PPP	HHH
Japan	NNN	NNN	PPP	PPP	PPP	HHH	HHH	PPP	PPP	PPP	HHH

Table 18 compares the results obtained using CGM and LFLA at the ontological process ONTA leading to OSR for two different values of orness (ω~0.3, 0.8), while assuming perfect (PPP) truth degrees (or levels of confidence). The CGM method predicts a perfect (PPP) ontological relationship between the three variables CRV, TSS and TCS and the outcome variable OSR, for all states considered in the study. This is in agreement with the LFLA method for low level of orness (ω~0.3) and through the LWD operation. This case of LFLA method also predicts a perfect (PPPP) level of confidence or truth. However, when we move to the case of LFLA with the LWA aggregation operator or for high level of orness (ω~ 0.8), we see much more variance across the eight states. One noticeable feature though is that France, Russia 1917 and China (i.e., the cases with a successful social revolution) there is no variation at all across the table – the LFLA method predicts the exact same results across all three cases. The case of LFLA with LWC with ω~ 0.8 is somewhat unique for we see in here that the model predicts only a high (HHH) nuanced ontological relationship for France, Russia 1917, and China, whereas it predicts a very high (VHH) nuanced ontological relationship for

Table 18 Comparing CGM and LFLA Analysis of Skocpol's Theory: Ontological Relationship ONTA Leading to OSR for ω~0.3, 0.8 and PPP input truth values.

				ONTA								
	CRV	TSS	TCS	CGM	LWC: ω~0.3		LWC: ω~0.8		LWA: ω~0.3		LWA: ω~0.8	
State	ν	ν	ν	ν	ν	τ	ν	τ	ν	τ	ν	τ
France	PPP	PPP	PPP	PPP	PPP	PPP	HHH	HHH	PPP	PPP	PPP	HHH
Russia 1917	PPP	PPP	PPP	PPP	PPP	PPP	HHH	HHH	PPP	PPP	PPP	HHH
China	PPP	PPP	PPP	PPP	PPP	PPP	HHH	HHH	PPP	PPP	PPP	HHH
England	NNN	PPP	LLL	PPP	PPP	PPP	VHH	VHH	VLL	VLL	VLL	VLL
Russia 1905	PPP	NNN	NNN	PPP	PPP	PPP	VHH	VHH	NNN	NNN	NNN	NNN
Germany	MMM	NNN	NNN	PPP	PPP	PPP	PPP	PPP	NNN	NNN	NNN	NNN
Prussia	NNN	LLL	MMM	PPP	PPP	PPP	PPP	PPP	VLL	VLL	VLL	VLL
Japan	NNN	PPP	LLL	PPP	PPP	PPP	VHH	VHH	VLL	VLL	VLL	VLL

England, Russia 1905, and Japan and a perfect (PPP) one for Germany and Prussia. Another noticeable feature of Table 18 is that the LFLA model using LWA predicts a null (NNN) ontological relationship for Russia 1905 and Germany. However, these last results are highly dubious since the model predicts a null (NNN) level of confidence in these estimates. In conclusion, it is safe to argue that the ontological relationship depicted by Skocpol's theory is empirical valid to a large extent, especially for France, Russia 1917 and China. Having said this, it is also important to note that one value-added of the LFLA method is that it shows much variance between the states wherein there was a successful social revolution and those where the revolutions did not succeed.

Table 19 compares the results obtained using CGM and LFLA at the basic causality process BCAUS leading to CSR for two different values of orness (ω~0.3, 0.8), while assuming perfect (PPP) truth degrees (or levels of confidence). The CGM method predicts a perfect (PPP) basic causal process between the two variables SBR and PRV and the outcome variable CSR, for all states considered in the study. This is in agreement with the LFLA method for low level of orness (ω~0.3) and through the LWC operation. This case of LFLA method also predicts a perfect (PPPP) level of confidence or truth. The three other cases of LFLA modeling shown in the table agree with this for France and Russia 1917 (if with a lower level – HHH – of nuanced causality in the case of LWC with ω~0.8). The China case stands apart from its homologues in the basic causality process when LWA is used in the LFLA method – the models predict a lower level (VHH) of nuanced causality for China. Another noticeable result of this table is that Germany and Prussia are predicted to have a PPP nuanced basic causality with PPP level of confidence when using LWC for ω~0.8 whereas the cases of France, Russia 1917, and China (those with successful social revolutions) are predicted to have only HHH nuanced causality with a HHH level of confidence.

What to make of all this variance in the predictions across different LFLA methods? First, it shows the richness of the modeling endeavor of linguistic fuzzy logic approach to studying multi-level causality – a richness that is quite appropriate for the task at hand as the empirical cases do indeed show a lot of diversity and

Table 19 LFLA Analyses of Skocpol's Theory: Basic Causal Process BCAUS Leading to CSR for ω~0.3 thru LWDC and LWA and PPP input truth values.

			BCAUS								
	SBR	PRV	CGM	LWC: ω~0.3		LWC: ω~0.8		LWA: ω~0.3		LWA: ω~0.8	
State	V	V	V	V	τ	V	τ	V	τ	V	τ
France	PPP	PPP	PPP	PPP	PPP	HHH	HHH	PPP	PPP	PPP	HHH
Russia 1917	PPP	PPP	PPP	PPP	PPP	HHH	HHH	PPP	PPP	PPP	HHH
China	PPP	HHH	PPP	PPP	PPP	HHH	HHH	VHH	VHH	VHH	HHH
England	PPP	NNN	PPP	PPP	PPP	VHH	VHH	MLL	MLL	HHH	MHH
Russia 1905	MMM	PPP	PPP	PPP	PPP	VHH	VHH	MHH	MHH	VHH	HHH
Germany	LLL	MMM	PPP	PPP	PPP	PPP	PPP	MLL	MLL	MLL	MLL
Prussia	HHH	MMM	PPP	PPP	PPP	PPP	PPP	MHH	MHH	MHH	MHH
Japan	HHH	NNN	PPP	PPP	PPP	VHH	VHH	LLL	LLL	MMM	MMM

Table 20 Effect of Input Truth Values: Results for Causality Processes for ω~ 0.3.

	PPP truth value for input variables											
	SCSBR thru LWD		SCPRV thru LWD		BCAUS thru LWC		SCSBR thru LWA		SCPRV thru LWA		BCAUS thru LWA	
State	V	τ	V	τ	V	τ	V	τ	V	τ	V	τ
France	PPP	PPP	PPP	PPP	PPP	PPP	PPP	PPP	PPP	PPP	PPP	PPP
Russia 1917	PPP	PPP	PPP	PPP	PPP	PPP	PPP	PPP	PPP	PPP	PPP	PPP
China	PPP	PPP	PPP	PPP	PPP	PPP	PPP	PPP	PPP	PPP	VHH	VHH
England	PPP	PPP	PPP	PPP	PPP	PPP	PPP	PPP	PPP	PPP	MLL	MLL
Russia 1905	MMM	MMM	PPP	PPP	PPP	PPP	MMM	MMM	PPP	PPP	MHH	MHH
Germany	LLL	HHH	PPP	PPP	PPP	PPP	LLL	LLL	PPP	PPP	MLL	MLL
Prussia	HHH	HHH	PPP	PPP	PPP	PPP	HHH	HHH	PPP	PPP	MHH	MHH
Japan	HHH	HHH	PPP	PPP	PPP	PPP	HHH	HHH	PPP	PPP	LLL	LLL

Table 21 Effect of Input Truth Values: Results for Causality Processes for ω~ 0.3.

	MHH truth value for input variables											
	SCSBR thru LWD		SCPRV thru LWD		BCAUS thru LWC		SCSBR thru LWA		SCPRV thru LWA		BCAUS thru LWA	
State	V	τ	V	τ	V	τ	V	τ	V	τ	V	τ
France	MHH	PPP	MHH	PPP	MHH	MHH	MHH	MHH	MHH	MHH	MHH	MHH
Russia 1917	MHH	PPP	MHH	PPP	MHH	MHH	MHH	MHH	MHH	MHH	MHH	MHH
China	MHH	PPP	MHH	PPP	MHH	MHH	MHH	MHH	MHH	MHH	MHH	MHH
England	MHH	PPP	MHH	PPP	MHH	MHH	MHH	MHH	MHH	MHH	MMM	MMM
Russia 1905	MMM	MMM	MHH	PPP	MHH	MHH	MMM	MMM	MHH	MHH	MMM	MMM
Germany	MLL	HHH	MHH	PPP	VHH	VHH	MLL	MLL	MHH	MHH	MLL	MLL
Prussia	MHH	HHH	MHH	PPP	HHH	HHH	MHH	MHH	MHH	MHH	MMM	MMM
Japan	MHH	HHH	MHH	PPP	MHH	MHH	MHH	MHH	MHH	MHH	MMM	MMM

variance, so much so that scholars have kept arguing since Skocpol wrote her book on how and to what extent her empirical data falsifies her theory (as neatly summarized by GM). Second, the use of a parameter – the level of orness ω – allows us some "testing" space for fitting different models derived from using

Table 22 Effect of Input Truth Values: Results for Causality Processes for $\omega\sim 0.3$.

	Uniformly randomized truth values for input variables											
	SCSBR thru LWD		SCPRV thru LWD		BCAUS thru LWC		SCSBR thru LWA		SCPRV thru LWA		BCAUS thru LWA	
State	ν	τ	ν	τ	ν	τ	ν	τ	ν	τ	ν	τ
France	PPP	PPP	MHH	PPP	VHH	MLL	PPP	PPP	MLL	MLL	VLL	VLL
Russia 1917	MMM	PPP	MHH	PPP	MHH	MLL	LLL	LLL	MLL	MLL	LLL	LLL
China	MHH	PPP	MHH	PPP	MHH	LLL	MHH	MHH	MLL	MLL	VLL	VLL
England	HHH	PPP	MHH	PPP	MHH	MHH	HHH	HHH	MLL	MLL	MHH	MHH
Russia 1905	MMM	MMM	MHH	PPP	MHH	HHH	MMM	MMM	MLL	MLL	MLL	MLL
Germany	HHH	HHH	MHH	PPP	VHH	VHH	MLL	MLL	MLL	MLL	VHH	VHH
Prussia	HHH	HHH	MHH	PPP	HHH	HHH	MLL	MLL	MLL	MLL	MHH	MHH
Japan	PPP	HHH	MHH	PPP	MHH	HHH	HHH	HHH	MLL	MLL	MHH	MHH

Table 23 Effect of Input Truth Values: Results for Ontological Process ONTA for $\omega\sim 0.3$

	PPP truth value for input variables				MHH truth value for input variables				Uniformly randomized truth values for input variables			
	thru LWC		thru LWA		thru LWC		thru LWA		thru LWC		thru LWA	
State	ν	τ	ν	τ	ν	τ	ν	τ	ν	τ	ν	τ
France	PPP	PPP	PPP	PPP	MHH	MHH	MHH	MHH	MHH	VLL	VLL	VLL
Russia 1917	PPP	PPP	PPP	PPP	MHH	MHH	MHH	MHH	MHH	MHH	MHH	MHH
China	PPP	PPP	PPP	PPP	MHH	MHH	MHH	MHH	MHH	MLL	MLL	MLL
England	PPP	PPP	VLL	VLL	MHH	MHH	MLL	MLL	MHH	MHH	MHH	MHH
Russia 1905	PPP	PPP	NNN	NNN	MHH	MHH	MLL	MLL	MHH	MHH	MHH	MHH
Germany	PPP	PPP	NNN	NNN	VHH	VHH	MLL	MLL	VHH	VHH	MMM	MMM
Prussia	PPP	PPP	VLL	VLL	HHH	HHH	MLL	MLL	HHH	HHH	HHH	HHH
Japan	PPP	PPP	VLL	VLL	MHH	MHH	MLL	MLL	MHH	MHH	MHH	MHH

Table 24 Sample of uniformly randomized truth values used for input variables

	Random Input τ							
State	INT	DCL	AGB	PAS	LAV	CRV	TSS	TCS
France	VLL	MMM	MLL	NNN	VHH	LLL	NNN	MHH
Russia 1917	VHH	HHH	NNN	NNN	VHH	MHH	MHH	HHH
China	MMM	MLL	HHH	VHH	VLL	VHH	MHH	LLL
England	MHH	LLL	MLL	MLL	HHH	HHH	NNN	MMM
Russia 1905	VHH	VHH	MHH	LLL	LLL	HHH	MLL	MLL
Germany	HHH	MMM	NNN	VHH	HHH	VHH	NNN	NNN
Prussia	NNN	HHH	VLL	LLL	HHH	VHH	MHH	VHH
Japan	VHH	NNN	MHH	VHH	VLL	VLL	VHH	HHH

LFLA. In other words, the fuzziness level of causality and ontological relations is not to be assumed as a pre-theoretical given. Rather, the LFLA method makes it an integral part of the investigation. Third, the richness of the LFLA method to studying causality and ontological relationships is further illustrated in Tables 20, 21, 22, and 23, where I compare the results obtained for low levels of fuzziness (or orness) at different values of the truth values of the input data. We can clearly see that what we assume for truth values for the input variables does indeed have an important impact on the levels of nuance and confidence of the various causal and ontological process of Skocpol's theory. This hence reinforces the call made earlier in the book for developing ways to systematically estimate this crucial aspect – i.e., the truth value or level of confidence – of the empirical data, an aspect of data analysis that has yet to be considered in social sciences.[7]

The linguistic fuzzy logic approach offers a much richer spectrum of possible aggregation operators. Most of these operations lead to conclusions that are very different from CGM's analysis (and of course Boolean logic methods). As explained in Chapter 2, the notion of orness stands for the notion that when we express ourselves in natural language vagueness is inherently present. Aggregating linguistic information does not eliminate this inherent vagueness and thus the very process of aggregation itself is best constructed by making it in itself built on vagueness. More specifically, speaking of LWA as having a given degree of orness means that the LWA is *more or less like* a disjunction or a conjunction – *LWA possesses more or less or-like (or and-like) behavior.*[8] A key claim in this chapter is that logic can also be conceptualized in linguistic fuzzy terms. This means for example that we do not strictly speaking have necessary or sufficient conditions, but we rather have more or less necessary and more or less sufficient conditions. This can be formalized through the notion of orness that the LWA operation is built on. Because we think and express our thoughts and discourses in natural language, vagueness is inherent in our thoughts and discourses, including social sciences discourses about causality. Linguistic fuzzy logic is one way to bring in tune the vagueness of our thought processes and discourses and social sciences, but in a mathematically rigorous and axiomatized way.

---

[7] I am not accusing the social sciences community of not being aware of this problem, that is, of not knowing that empirical estimates of variables are always done to a certain level of confidence only. I am rather calling for a systematization of this concern and its inclusion in our methods of data analysis as a separate factor, rather than as part of the "global error" term.

[8] I should point out (as a reminder) that the process of aggregation through LWA is not arbitrary as it might seem. As explained in Chapter 2, I use the MAXENT principle in conjunction with orness to specify the LOWA process of aggregating linguistic information. MAXENT stipulates that in the process of making inferences based on incomplete information, we should select the probability distribution function with maximum entropy value given the observed data. The distribution function obtained using MAXENT provides the most conservative estimation of the unknown underlying distribution function. This choice singles out the most significant and least biased distribution and the one which best represents the true distribution. Using MAXENT in the aggregation process of linguistic fuzzy information is suitable since it falls well within the spirit of preserving as much vagueness and uncertainty as possible. When supplemented with the measure of *orness* we have a procedure which preserves the vagueness of the information as well as the multi-dimensionality character of the aggregation process.

Chapter 7
Linguistic Fuzzy-Logic Data Analysis

Data analysis is at the core of much of what counts as social sciences today. Yet, the tools that have been designed for these purposes remain rather limited in social sciences, and more so in political science and International Relations. For the most part scholars use statistical means to analyze data. However, as succinctly put by Braumoeller (2003:209), "theories that posit complex causation, or multiple causal paths, pervade the study of politics but have yet to find accurate statistical expression … To date, however, no one has made a concerted effort to describe how the empirical implications of theoretical models that posit causal complexity could be captured by statistical methods." Without belittling the gap in statistical methodology that Braumoeller highlights, there is a need for other types of tools – non-statistical approaches – that could be used to address issues that statistical analyses cannot reach such as conceptual vagueness.

Whether it is statistical/probabilistic or formal/mathematical modeling in nature the analysis of data assumes a given "logic" and a correspondingly underlying algebraic structure of this logic. Explicating the logic underpinning data analysis and the underlying algebra of this logic helps clearly explicate the fundamental assumptions that define the analysis. Most data analyses in social sciences assume the Boolean logic as the underlying logic with its associated two-valued algebraic structure. For example, Braumoeller (2003), in his promising approach to complex causation, clearly positions his method as being Boolean. Yet, there are no inherent theoretically, empirically, or methodologically insurmountable impediments or *raison d'être* why we cannot explore the vast realm of other logics in social science inquiry.

In this chapter, I specifically intend to do so – I propose a non-Boolean, non-statistical approach to study the logic of data analysis. This of course is not a call against statistical methods which have taught us a great deal about causality and more generally relations between dependent and independent variables. I also advocate "computing with words" as another way of doing quantitative analysis by adopting *linguistic* fuzzy logic as a framework. Although conventional numericalization of knowledge in social sciences has achieved much by seeking more sophistication and mathematical rigor, the latter can also be achieved but with words expressed in natural language, not numbers. That is, quantitative analysis is possible without assuming a Boolean logic and/or statistical analysis using the technology of "computing with words."

By way of introduction, I first delineate and discuss the commonalities and the most important differences between Ragin's (2000) approach and LFLA, thereby

B. Arfi: Linguistic Fuzzy Logic Methods in Social Sciences, STUDFUZZ 253, pp. 155–173.
springerlink.com

highlighting the value-added for the latter. I next briefly summarize Charles Ragin's approach to data analysis and the study of causality based on fuzzy-set theory. I then elaborate the discussion, therewith highlighting key overlaps and most important differences between Ragin's approach and LFLA. As a further illustration I also apply LFLA analysis to Braumoeller's (2003) re-examination of Huth's (1996) theory of war initiation.

I often refer to Charles Ragin's pioneering work on using fuzzy-set theory to analyze causality in social sciences for two reasons. First, this helps to set the ground for an understanding of the importance of using linguistic fuzzy logic to analyze causality in social sciences. Second, it allows me to highlight key overlaps between Ragin's approach and mine as well as the value-added of using linguistic fuzzy logic instead of just fuzzy set theory the way he does it. However, it is worthwhile keeping in mind that fuzzy sets are the essential elements of both approaches, the difference lying in the fact that I concentrate more on formal logic and that I use linguistic variables instead of membership functions – that is, numerical degrees of memberships to fuzzy sets – as Ragin does.

Two additional differences will also be discussed below. First, LFLA allows a fuzzification of the degrees of membership to a certain category. That is: *membership degrees are also considered to be fuzzy sets*, not crisp numbers as assumed by Ragin's approach. This is done through the concept of truth value as explicated down below. This allows the introduction of a fuzzy-logic equivalent of the confidence level used in statistical studies. Second, LFLA allows a fuzzification of the dichotomous division into necessary and sufficient conditions, that is, the notion of simple causality is fuzzified. Fuzzy relations of necessity or sufficiency are expressed as having a certain degree of fuzzy nuance with a degree of truth. We would, for example, say that there is very high fuzzy relation of necessity with a low degree of truth.

1 Linguistic Fuzzy-Logic Data Analysis of Simple Causality

Simple causality stands for causal necessity, causal sufficiency, or a combination of both types of causal conditions. This is summarized in Table 1.

Table 1 Different Types of Simple Causation

Simple Causality	Conventional Semantic	Formal Logic	Alternative Form
Necessity	***Only if X then Y:*** *X necessarily causes Y*	$Y \to X$	$\neg X \to \neg Y$
Sufficiency	***If X then Y:*** *X sufficiently causes Y*	$X \to Y$	$\neg Y \to \neg X$
Necessity and Sufficiency	***If and Only if X then Y:*** *X necessarily and sufficiently causes Y*	$(X \to Y) \wedge (Y \to X)$	$(X \to Y) \wedge (\neg X \to \neg Y)$

Let us start with necessary causation. X necessarily causes Y means that X is an antecedent condition for all positive outcomes of Y. Put in terms of fuzzy set theory: Ragin (2000:272) phrases the question as "Do instances of the outcome constitute a subset of one or more causal conditions?" Translated in terms of membership scores this means that "If fuzzy-membership scores in the outcome are uniformly less than or equal to fuzzy-membership scores in the cause (yielding a 'lower triangular' scatterplot), then the cause may be considered a necessary condition for the outcome" (2000:272). Ragin combines this with a probabilistic test of necessity with "almost necessary" (~0.80) as the benchmark and 0.01 as the significance level to explore necessary causation in his study of IMF protest.[1] He finds that IMF pressure (IMP) and Urbanized (URB) pass this test of necessity.

The starting point of the analysis of necessity in linguistic fuzzy logic approach is that we have a logical inference such as Y → X with X being tested as a necessary condition for Y.[2] Let us phrase this using the language of linguistic fuzzy logic:

1. *We know that: S1={y has nuance value Y with a degree of truth* τ_y*}*
2. *We know that: S2={x has nuance value X with a degree of truth* τ_x*}*
3. *To what degree of nuanced (fuzzy) necessity* V_n *and degree of truth* τ_n *can we deduce that {If S1 then S2}?*

That is: We know the linguistic nuance value *Y* of *y* and its truth value τ_y. We know the linguistic nuance value *X* of *x* and its truth value τ_x. Can we infer from this the nuance value V_n and truth value τ_n of the fuzzy statement "**X necessarily causes Y**"? In solving this problem we need therefore to evaluate the nuance and truth values of the logical inference using the linguistic implication operator, LI. As an illustration consider the following query on whether we have IMP (IMF Pressure) as a necessary condition for PRO (IMF Protest): PRO → IMP? For example, as shown in Chapter 2, in the Argentinean case we have the nuance values which correspond to Ragin's degrees of membership: $v_{PRO} = PPP$ (perfectly included in the category of IMF protest) and $v_{IMP} = HHH$ (highly included in the category of IMF Pressure). However, this information is not enough since in the LFLA we also need to know the degrees of truth τ_{PRO} and τ_{IMP} of these attributions of degrees of membership (which correspond to what I call nuance values). Using the linguistic implication operation we can thus write that:

$$\left(v_{PRO\to IMP},\tau_{PRO\to IMP}\right)=LI\left[\left(v_{PRO},\tau_{PRO}\right),\left(v_{IMP},\tau_{IMP}\right)\right]=LWD\left[Neg\left\{\left(v_{PRO},\tau_{PRO}\right)\right\};\left(v_{IMP},\tau_{IMP}\right)\right] \quad (157)$$

This leads more explicitly to the following expressions for the degree of nuanced necessity $v_{PRO\to IMP}$ and its corresponding degree of truth $\tau_{PRO\to IMP}$:

---

[1] See Chapter 2 for a summary of relevant Ragin's analysis of IMF protest.

[2] Note that we could equivalently use the alternative form: ¬ **X** → ¬ **Y**, that is, a non-occurrence of X implies a non-occurrence of Y.

$$v_{PRO \to IMP} = MAX\left\{MIN\left[Neg\left(v_{PRO}, \tau_{PRO}\right)\right]; MIN\left[\left(v_{IMP}, \tau_{IMP}\right)\right]\right\} \tag{158}$$

$$\tau_{PRO \to IMP} = LOWA\left(Neg\left(\tau_{PRO}\right), \tau_{IMP}\right) \tag{159}$$

Assuming for the sake of illustration that the degrees of truth in the case of Argentina[3] are: $\tau_{PRO} = VHH$ and $\tau_{IMP} = MMM$. That is: the degree of membership of Argentina in the set of IMF Protest is known to very high degree of truth (or confidence) and its membership in the set of IMF Pressure is known to a moderate degree of truth (or confidence). We also know from Ragin's data that $v_{PRO} = PPP$ and $v_{IMP} = HHH$. Plugging these nuance and truth values into the formula we get:

$$v_{PRO \to IMP} = MAX\left\{MIN\left[Neg\left(PPP, VHH\right)\right]; MIN\left[\left(HHH, MMM\right)\right]\right\} \tag{160}$$

$$\tau_{PRO \to IMP} = LOWA\left(Neg\left(VHH\right), MMM\right) \tag{161}$$

Using the definitions of Neg, MAX, MIN, and LOWA operators (see Chapter 2) we obtain:

$$v_{PRO \to IMP} = MAX\left\{MIN\left[\left(NNN, VLL\right)\right]; MIN\left[\left(HHH, MMM\right)\right]\right\} = MAX\left\{NNN; MMM\right\} = MMM \tag{162}$$

$$\tau_{PRO \to IMP} = LOWA\left(Neg\left(VHH\right), MMM\right) = LOWA\left(VLL, MMM\right) = MMM \tag{163}$$

That is: IMF Pressure (IMP) is a moderately (MMM) necessary condition for IMF Protest (PRO) to a moderate (MMM) degree of truth. Note that had we assumed that the nuances of PRO and IMP were perfectly known with $\tau_{PRO} = PPP$ and $\tau_{IMP} = PPP$, we would have had $\tau_{PRO \to IMP} = PPP$, that is, we would be able to say with a perfect degree of truth (or level of confidence) that IMF Pressure (IMP) is a moderately necessary condition for IMF Protest (PRO). In Ragin's approach, the level of confidence in the inference is addressed through the practice of "benchmarking" in combination with a null hypothesis testing at a preset level of significance; that is, combining fuzzy-set approach with a conventional statistical test of significance. *The linguistic fuzzy logic approach is thus (comparatively speaking) self-sufficient* (for the lack of a better term): the confidence in the

[3] These are unfortunately not given in Ragin's empirical research.

inference of whether we have a fuzzy necessary condition is an internal byproduct of the same method. Both the level of nuanced causation and the confidence level (truth value) of this estimate are simultaneously produced without resorting in addition to a statistical procedure.

Furthermore, not only a sort of benchmarking is inherent in the method itself through the concept of degree of truth. The method also introduces a sort of individual benchmarking (i.e., individual truth levels) for each of the variables (or categories) being considered in the analysis. In comparison, Ragin's method uses one overall procedure of benchmarking with a preset level of confidence.[4] Consider as an example the fuzzy set (category) IMF pressure (IMP). Ragin computes the proportion of countries that are included in the fuzzy set using the criterion:

$$Whether\left\{\mu_{IMP}\left(C_i\right) \geq \mu_{PRO}\left(C_i\right)\right\} \text{ for country } C_i \tag{164}$$

If we denote by N the total number of countries in the study, the proportion for a variable (fuzzy set) X is given by:

$$P_X = \frac{1}{N}\sum_{j=1}^{N}\left\{1-\Theta\left[\mu_{PRO}\left(C_j\right)-\mu_X\left(C_j\right)\right]\right\} \text{ with } \Theta(a)=\begin{cases}1 & if \quad a>0 \\ 0 & if \quad a\leq 0\end{cases} \tag{165}$$

Ragin finds for example that $P_{IMP} = 94\%$ (see his Table 10.2, p. 273). If p denotes the benchmark proportion, Ragin (2000:111) then uses (for N > 30, otherwise he would use a binomial distribution) the following z-score to evaluate the confidence level of the procedure (through null hypothesis testing with a preset significance level of α):

$$z_X = \frac{\left(P_X - p\right) - \dfrac{1}{2N}}{\sqrt{\dfrac{p(1-p)}{N}}} \tag{166}$$

Ragin finds that two variables – IMP and URB – pass the test of necessity with a benchmark of $p = 0.80$ and a significance level of $\alpha = 0.01$. What is the equivalent procedure in linguistic fuzzy logic? The answer lies in using the operator LWA (linguistic weighted average) in conjunction with the operator LI (linguistic implication), both introduced in Chapter 2.

[4] Obviously, this strength of the LFLA method comes at a price – in addition to the nuance degrees (or equivalently degrees of membership) we need also to empirically estimate the degrees of truth of these nuances. Addressing this issue is an important aspect of empirical testing but is left for future work as far as this book is concerned. Note that this issue has been addressed to a certain extent in the artificial intelligence literature on type-2 fuzzy sets, if however only for numerical membership functions, not linguistically expressed ones.

First, we need to recast the criterion: $Whether\left\{\mu_{IMP}\left(C_i\right)\geq\mu_{PRO}\left(C_i\right)\right\}$ using the LI operation since, as explained earlier, the criterion is effectively a numerical translation of the formal logical implication {PRO→IMP}. In Ragin's approach the above criterion lends either 0 or 1/N through the formula $\left\{1-\Theta\left[\mu_{PRO}\left(C_j\right)-\mu_X\left(C_j\right)\right]\right\}/N$, with the sum of these elements giving the proportion for the variable X. In the LFLA method we obtain a nuance degree and a truth degree through the formula:

$$\left(v_{PRO\to IMP}\left(C_j\right),\tau_{PRO\to IMP}\left(C_j\right)\right)=LI_{\omega}\left\{\left(v_{IMP}\left(C_j\right),\tau_{IMP}\left(C_j\right)\right);\left(v_{PRO}\left(C_j\right),\tau_{PRO}\left(C_j\right)\right)\right\} \tag{167}$$

Second, the equivalent of the proportion is obtained using the aggregation operation LWA as:

$$\left(v_{IMP}^{fnec}\left(\omega\right),\tau_{IMP}^{fnec}\left(\omega\right)\right)=LWA_{\omega}\left\{\left(v_{PRO\to IMP}\left(C_1\right),\tau_{PRO\to IMP}\left(C_1\right)\right);\ldots;\left(v_{PRO\to IMP}\left(C_N\right),\tau_{PRO\to IMP}\left(C_N\right)\right)\right\} \tag{168}$$

Using the definition of LWA (see Chapter 2) we see that:

$$v_{IMP}^{fnec}\left(\omega\right)=LOWA_{\omega}\left\{v_{PRO\to IMP}\left(C_1\right);\ldots;v_{PRO\to IMP}\left(C_N\right)\right\} \tag{169}$$

$$\tau_{IMP}^{fnec}\left(\omega\right)=LOWA_{\omega}\left\{MIN\left(v_{PRO\to IMP}\left(C_1\right),\tau_{PRO\to IMP}\left(C_1\right)\right);\ldots;MIN\left(v_{PRO\to IMP}\left(C_N\right),\tau_{PRO\to IMP}\left(C_N\right)\right)\right\} \tag{170}$$

$v_{IMP}^{fnec}\left(\omega\right)$ and $\tau_{IMP}^{fnec}\left(\omega\right)$ are respectively the nuance of fuzzy necessity and the degree of truth (or confidence) of this fuzzy necessity (with a degree of orness ω as explicated in Chapter 2). The concept of fuzzy necessity generalizes the notion of crisp necessity as usually thought of in social sciences. Fuzzy necessity with varying levels of nuance corresponds to more or less necessity, which can be roughly expressed as some necessity contaminated with some contingency. Hence, fuzzy necessity (or, conversely, fuzzy contingency) varies on a spectrum extending from crisp necessity to crisp contingency.

An important question is: What is the interpretation of the degree (or level) orness ω which is embedded in the evaluation of LWA_{ω} through the $LOWA_{\omega}$ operation? As explained in Chapter 2, orness ω is a measure of the extent to which the aggregation of linguistic information is or-like, that is, whether we are closer to an intersection or union of the fuzzy sets representing the different categories. An orness ω ~1 represents a crisp intersection of the fuzzy sets, thereby corresponding to the smallest common fuzzy set which is included in all fuzzy sets. An orness ω ~ 0 represents a crisp union of the fuzzy sets, thereby corresponding to the largest of the fuzzy sets which contains all fuzzy sets. Hence, for very small degrees of orness (thus high degrees of andness) we move toward recovering crisp necessity, whereas for high degrees of orness (thus small degrees of andness) we move toward recovering crisp contingency.

Ragin's concept of benchmark p is such that $P_X \geq p$ for all cases that pass the threshold test (set by the benchmark), thereby corresponding to an inclusion of all these fuzzy categories X. Ragin relies on linguistic qualifiers of the benchmark such as "most often than not" ($p = 0.5$), "usually" ($p = 0.65$), "almost always" ($p = 0.80$). The procedure of benchmarking is probabilistically evaluated through the notion of significance level. Yet it is a crisp – not fuzzy – notion as shown in equation [165] for the computation of proportions. In comparison, the LWA_ω procedure fuzzifies this benchmarking since the weights resulting from using the MAXENT (maximum entropy) principle are efficiently distributed in such a way as to best represent the existing information. For the crisp benchmarking method the weights are either 0 or 1/N meaning that a case is "in" with a weight 1/N or is "out", thereby showing how Boolean logic still indirectly determines Ragin's fuzzy-set approach to the study of causality. In the LWA approach all cases contribute more or less, depending on the level of andness (or orness ω).

To test these conclusions, I conduct a hypothetical experiment where all cases contribute perfectly to all fuzzy variables. In Ragin's method this would mean that all proportions are exactly 1 for all independent variables. In the LFLA method it means that $v_{X_k}(C_i) = PPP$ for all variables X and all countries C_i. I also assume that the truth degrees of all these attributions are perfect, $\tau_{X_k}(C_i) = PPP$. The expectation is to find perfect necessity. Doing the calculations leads to $v_{X_k}^{fnec} = PPP$ and $\tau_{X_k}^{fnec} = PPP$ for all variables X for a degree of andness of 0.999 (orness $\omega = 0.001$) – we indeed have perfect necessity with a perfect degree of confidence (truth), as it indeed should be.

Applying the same method to Ragin's actual data for the nuances of all variables gives different results depending, first, on the level of orness ω and, second, on what one assumes about the confidence (truth) of Ragin's assignment of these nuances to the various variables in the 35 countries that he considered. The expectation however is that the smaller the orness level and the higher the truth degrees assumed for Ragin's raw data, the closer the results of LFLA should be to Ragin's fuzzy-set theory combined with statistical method. The results are shown in Table 2.

Ragin's conclusions about IMP and URB are confirmed, if with a number of important differences. First, the approach simultaneously provides the level of fuzzy necessity relation and the level of truth of this estimate (which can also be interpreted as the level of confidence). Second, the LFLA method shows a difference between the two variables IMP and URB. Whereas Ragin finds the same confidence for both these variables, LFLA shows using the same original Ragin's data that IMP (IMF pressure) is less necessary than URB (urbanization) – IMP displays a very low fuzzy necessity whereas URB displays a low fuzzy necessity. Moreover, URB is estimated with a low degree of confidence, which is higher than that of IMP which is estimated as very low.

Table 2 LFLA approach to Ragin's data with an orness level of ω = 0.001 assuming a perfect confidence (PPP) in the measured data

	ω = 0.001 with PPP empirical confidence		ω = 0.001 with HHH empirical confidence		ω = 0.6 with PPP empirical confidence	
Variable	$\nu_{X_k}^{fnec}$	$\tau_{X_k}^{fnec}$	$\nu_{X_k}^{fnec}$	$\tau_{X_k}^{fnec}$	$\nu_{X_k}^{fnec}$	$\tau_{X_k}^{fnec}$
IMP	VLL	VLL	LLL	LLL	VLL	VLL
URB	LLL	LLL	LLL	LLL	LLL	LLL
ECH	NNN	NNN	LLL	LLL	NNN	NNN
DEI	NNN	NNN	LLL	LLL	NNN	NNN
POL	NNN	NNN	LLL	LLL	NNN	NNN
GOA	NNN	NNN	LLL	LLL	VLL	VLL

These results will change if one assumes a lower degree of confidence in Ragin's raw data. If instead of assuming a perfect confidence (PPP) I assume, say a high confidence (HHH) only, keeping the same level of orness (ω=0.001), we obtain the results in Table 2. We clearly see that all variables show the same low nuance of fuzzy necessity at the same low level of confidence. If instead we keep the same level confidence in the raw data, that is, HHH, but change the orness level to $\omega = 0.1$, 0.3, 0.8, or 0.95 we obtain the same results (not shown in Table 2). However, keeping a PPP level of confidence in Ragin's raw data but increasing the level of orness, for example, to $\omega = 0.6$, produces the results in Table 2, which shows that GOA (government activism) behaves now in the same way as IMP while no change occurs for the other variables.

In sum, LFLA method does confirm Ragin's results obtained in using a combination of fuzzy-set theory and statistical null hypothesis testing with benchmarking. In addition, LFLA method qualifies Ragin's method by also showing some hidden assumptions concerning the confidence level of the raw data (that is, Ragin implicitly assumes a perfect confidence). Moreover, the LFLA method avoids any resort to statistical testing as a way of validating its results. Finally, necessity inferred from empirical data is shown to be a fuzzy notion with a nuance level and a truth degree that can vary depending on the nuance levels and truth degrees of raw data as well as the level of orness that one opts for. Of course, much of this discussion is equivalently applicable to the study of sufficiency as well as conjoint instances of necessity and sufficiency. In the following section I compare LFLA to Braumoeller's (2003) Boolean logit study of complex causality as he applied it to reexamine Huth's (1996) exploration of the initiation of international wars.

2 Linguistic Fuzzy Logic Data Analysis of Multiple-Path Causation

In an innovative study Braumoeller (2003:210, 216, 219) suggests a sophisticated statistical method – Boolean logit/probit – to test theories that posit multiple paths resulting from conjectural causation and substitutability to (non)outcome. Conjectural causation occurs when for example two independent variables x_1 and x_2 (conjecturally) cause y. Denoting the corresponding probabilities respectively by p_{x_1} and p_{x_2}, p_y, the probability of the outcome, is:

$$p_y = p_{x_1} \times p_{x_2} \tag{171}$$

Substitutability occurs when two independent variables x_1 and x_2 can be substituted for one another to produce y. The probability for such an outcome is:

$$p_y = 1 - \left[\left(1 - p_{x_1}\right) \times \left(1 - p_{x_2}\right)\right] \tag{172}$$

The formulae of formal logic corresponding to these conditions are:

- Conjectural causation:

$$(X_1 \rightarrow Y) \wedge (X_2 \rightarrow Y) \tag{173}$$

- Substitutability:

$$\neg\left\{\left(\neg(X_1 \rightarrow Y)\right) \wedge \left(\neg(X_2 \rightarrow Y)\right)\right\} \text{ or equivalently } (X_1 \rightarrow Y) \vee (X_2 \rightarrow Y) \tag{174}$$

Braumoeller converts these probabilities into the framework of likelihood functions so as to use MLE (maximum likelihood estimation) to evaluate the statistical parameters of the causal model. He uses the following correspondence schema:

- Conjectural causation:

$$\begin{array}{ccccc} p_y & = & p_{x_1} & \times & p_{x_2} \\ \downarrow & & \downarrow & & \downarrow \\ \Pr\left(y_i = 1 \middle| \alpha_1, \beta_1, \alpha_2, \beta_2, x_i\right) & = & \Phi\left(\alpha_1 + \beta_1 x_{i1}\right) & \times & \Phi\left(\alpha_2 + \beta_2 x_{i2}\right) \end{array} \tag{175}$$

$\Phi(.)$ is a cumulative distribution function. This expression is generalized to J causal paths as:

$$\Pr\left(z_i = 1 \middle| \alpha, \beta, x_i\right) = \prod_{j=1}^{J} \left[\Phi\left(\alpha_j + \beta_j x_{ij}\right)\right] \tag{176}$$

The likelihood function is:

$$L(Y|\alpha,\beta,X)=\prod_{i=1}^{N}\left(\prod_{j=1}^{J}\left[\Phi\left(\alpha_j+\beta_j x_{ij}\right)\right]\right)^{y_i}\left(1-\prod_{j=1}^{J}\left[\Phi\left(\alpha_j+\beta_j x_{ij}\right)\right]\right)^{1-y_i} \tag{177}$$

For K independent variables per causal path, the likelihood function for conjectural causation (and a corresponding one for substitutability) becomes:

$$L(Y|\alpha,\beta,X)=\prod_{i=1}^{N}\left(\prod_{j=1}^{J}\left[\Phi\left(\alpha_j+\sum_{k=1}^{K}\beta_{jk}x_{ijk}\right)\right]\right)^{y_i}\left(1-\prod_{j=1}^{J}\left[\Phi\left(\alpha_j+\sum_{k=1}^{K}\beta_{jk}x_{ijk}\right)\right]\right)^{1-y_i} \tag{178}$$

The MLE method then gives the parameters α and β. What is the corresponding procedure in the LFLA method? Let us first express conjectural causation and substitutability in the LFLA formalism. Conjectural causation can be phrased using the language of linguistic fuzzy logic as:

- *We know that: X has nuance* v_X *to a degree of truth* τ_X
- *We know that: Y has nuance* v_Y *to a degree of truth* τ_Y
- *We know that: Z has nuance* v_Z *to a degree of truth* τ_Z
- *To what nuance of conjectural causation* v_{conj} *and degree of truth* τ_{conj} *can we deduce that: [{X causes Z} AND {Y causes Z}]?*

Substitutability can be phrased using the language of linguistic fuzzy logic as:

- *We know that: X is* v_X *to a degree of truth* τ_X
- *We know that: Y is* v_Y *to a degree of truth* τ_Y
- *We know that: Z is* v_Z *to a degree of truth* τ_Z
- *To what nuance of substitutability* v_{sub} *and degree of truth* τ_{sub} *can we deduce that: [{X causes Z} OR {Y causes Z}]?*

Using the LI, and LWC, and LWD operations, we can write for the nuance and truth values:

- Conjectural causation:

$$\left(v_{conj}(\omega),\tau_{conj}(\omega)\right)=LWC_{\omega}\left\{LI_{\omega}\left\{(v_X,\tau_X);(v_Z,\tau_Z)\right\};LI_{\omega}\left\{(v_Y,\tau_Y);(v_Z,\tau_Z)\right\}\right\} \tag{179}$$

- Substitutability:

$$\left(v_{sub}(\omega),\tau_{sub}(\omega)\right)=LWD_{\omega}\left\{LI_{\omega}\left\{(v_X,\tau_X);(v_Z,\tau_Z)\right\};LI_{\omega}\left\{(v_Y,\tau_Y);(v_Z,\tau_Z)\right\}\right\} \tag{180}$$

In Braumoeller's framework, the probability for an outcome O resulting from a combination of conjectural causation and substitutability of three variables A, B, and C takes the form:

$$p_O = p_A \times \left(1 - \left[(1 - p_B) \times (1 - p_C)\right]\right) \tag{181}$$

This multiple-path causation can be expressed in formal logic as:

$$(A \to O) \wedge \left[(B \to O) \vee (C \to O)\right] \tag{182}$$

This translates in LFLA formalism to:

$$\left(v_{mpc}(\omega), \tau_{mpc}(\omega)\right) = LWC_\omega\left\{\left(v_{A\to O}, \tau_{A\to O}\right); LWD_\omega\left\{\left(v_{B\to O}, \tau_{B\to O}\right); \left(v_{C\to O}, \tau_{C\to O}\right)\right\}\right\} \tag{183}$$

Where the nuance and truth levels of $X \to Y$ are given by:

$$\left(v_{X\to Y}, \tau_{X\to Y}\right) = LI_\omega\left\{\left(v_X, \tau_X\right); \left(v_Y, \tau_Y\right)\right\} \tag{184}$$

Note that the nuance and truth levels of multiple-path causation (equation [183]) depend on the level of orness ω.

As discussed earlier, LFLA enriches the discussion of causation further by fuzzifying the distinction between necessity and sufficiency, or, equivalently, conjectural causation and substitutability. Indeed, linguistic conjunction LWC_ω and disjunction LWD_ω are two limiting cases of the more general linguistic weighted averaging operation LWA_ω. The latter introduces another level of fuzziness that models fuzzy conjectural-like and fuzzy substitutability-like causation. The most general LFLA expression for multiple-path causality thus takes the form:

$$\left(v_{mpc}(\omega), \tau_{mpc}(\omega)\right) = LWA_\omega\left\{\left(v_{A\to O}, \tau_{A\to O}\right); LWA_\omega\left\{\left(v_{B\to O}, \tau_{B\to O}\right); \left(v_{C\to O}, \tau_{C\to O}\right)\right\}\right\} \tag{185}$$

Braumoeller uses his Boolean logit method to re-examine Huth's (1996) study of the puzzle of capabilities, resolve, and conflict initiation in international politics. In doing so, Braumoeller groups Huth's independent variables in three categories: (1) *capabilities* (CAP) which consists of one independent variable – balance of military forces (MBF); (2) *systemic incentives* (SIC) which consists of five independent variables – strategic location of territory (SLT), economic value of territory (EVT), prior gain of territory (PGT), shared alliance (SAL), and previous settlement (PST); and (3) *domestic incentives* (DIC) which consists of six variables – ties to bordering minority (TBM), political unification (PUN), economic value of territory (EVT), prior unresolved dispute (PUD), prior loss of territory (PLT), and decolonization norm (DNO). To compare the results obtained using LFLA to Braumoeller's method I use Huth's (1996) original data, which I convert to a linguistic form. Ten of the above eleven independent variables are dichotomous (coded 1 or 0) and thus are converted to either PPP or NNN. The variable MBF is a continuous one assuming values between 0 and 1. I convert it to a linguistic form using the scheme discussed in Appendix A.

LFLA requires two sorts of input data: nuance values and truth values. Huth's data provides only nuance values. As already pointed out in the comparison with Ragin's study of IMF protest, Huth's (and Braumoeller's) approach implicitly assumes that all truth values are 1 (i.e., PPP). This is not surprising since in Boolean logic nuance and truth values are in a one-to-one correspondence, that is, a measured nuance value always corresponds to a truth value of 1 (PPP in LFLA). To make the comparison with Huth's and Braumoeller's approaches meaningful I first assume that all truth values of the independent and dependent variables are PPP (1 in Boolean logic). I then relax this assumption to explore the impact of other values of truth on the overall conclusions of multiple-path analysis.

Explicitly applied to reexamine Braumoeller's approach to Huth's data, equation [182] can be broken in the following stages with DEV denoting Huth's dependent variable:

$$(MBF \rightarrow DEV) \wedge \left[(SIC \rightarrow O) \vee (DIC \rightarrow O)\right] \tag{186}$$

With *SIC* and *DIC* respectively standing for:

$$SIC \equiv SAL \wedge SLT \wedge PST \wedge EVT \wedge PGT \tag{187}$$

$$DIC \equiv PUN \wedge TMB \wedge DNO \wedge PUD \wedge EVT \wedge PLT \tag{188}$$

Expressed in LFLA formalism we obtain:

$$\left(v_{SIC \rightarrow DEV}; \tau_{SIC \rightarrow DEV}\right) = LWC\left\{\prod_{\substack{X = SAL, SLT, \\ PST, EVT, PGT}} \left(v_X; \tau_X\right)\right\} \tag{189}$$

$$\left(v_{DIC}; \tau_{DIC}\right) = LWC\left\{\prod_{\substack{X = PUN, TBM, \\ DNO, PUD, EVT, PLT}} \left(v_X; \tau_X\right)\right\} \tag{190}$$

Where:

$$\left\{\prod_{X = X_1, \ldots X_n} \left(v_X; \tau_X\right)\right\} = \left\{\left(v_{X_1}; \tau_{X_1}\right); \ldots; \left(v_{X_n}; \tau_{X_n}\right)\right\} \tag{191}$$

The nuance and truth levels for variable X for country C_j are obtained as previously discussed in the section dealing with the IMF protest as (with X = SIC or DIC):

$$\left(v_{X\to DEV}\left(C_j\right),\tau_{X\to DEV}\left(C_j\right)\right)=LI\left\{\left(v_X\left(C_j\right),\tau_X\left(C_j\right)\right);\left(v_{DEV}\left(C_j\right),\tau_{DEV}\left(C_j\right)\right)\right\} \tag{192}$$

Aggregation over all cases using LWA gives for variable X = SIC or DIC:

$$\left(v_{X\to DEV},\tau_{X\to DEV}\right)=LWA\left\{\left(v_{X\to DEV}\left(C_1\right),\tau_{X\to DEV}\left(C_1\right)\right);\ldots;\left(v_{X\to DEV}\left(C_N\right),\tau_{X\to DEV}\left(C_N\right)\right)\right\} \tag{193}$$

N is the number of cases. The nuance and truth levels for multiple-path causation are given by:

$$\left(v_{mpc},\tau_{mpc}\right)=LWC\left\{\left(v_{MBF\to DEV},\tau_{MBF\to DEV}\right);LWD\left\{\left(v_{DIC\to DEV},\tau_{DIC\to DEV}\right);\left(v_{SIC\to DEV},\tau_{SIC\to DEV}\right)\right\}\right\} \tag{194}$$

We can generalize the latter equation using the notion of fuzzy causality as discussed earlier (that is, generalizing from linguistic conjunction LWD and disjunction LWC to linguistic weighted averaging by using LWA) to obtain:

$$\left(v_{mpc},\tau_{mpc}\right)=LWA\left\{\left(v_{MBF\to DEV},\tau_{MBF\to DEV}\right);LWA\left\{\left(v_{DIC\to DEV},\tau_{DIC\to DEV}\right);\left(v_{SIC\to DEV},\tau_{SIC\to DEV}\right)\right\}\right\} \tag{195}$$

Before comparing LFLA and Boolean logit results for multi-path causation, I present first the results using LFLA without multi-path causation. In other words, I present the results of an analysis following the procedure that I introduced when comparing LFLA to Ragin's fuzzy-set based approach to causality. Recasting equations[167],[168], [169], and [170], we obtain for independent variable IND fuzzy-necessarily causing dependent variable DEV for the case C_j:

$$\left(v_{DEV\to IND}\left(C_j\right),\tau_{DEV\to IND}\left(C_j\right)\right)=LI_{\omega}\left\{\left(v_{IND}\left(C_j\right),\tau_{IND}\left(C_j\right)\right);\left(v_{DEV}\left(C_j\right),\tau_{DEV}\left(C_j\right)\right)\right\} \tag{196}$$

Thus for N cases:

$$\left(v_{IND}^{fnec}\left(\omega\right),\tau_{IND}^{fnec}\left(\omega\right)\right)=LWA_{\omega}\left\{\left(v_{DEV\to IND}\left(C_1\right),\tau_{DEV\to IND}\left(C_1\right)\right);\ldots;\left(v_{DEV\to IND}\left(C_N\right),\tau_{DEV\to IND}\left(C_N\right)\right)\right\} \tag{197}$$

Using the definition of LWA:

$$v_{IND}^{fnec}\left(\omega\right)=LOWA_{\omega}\left\{v_{DEV\to IND}\left(C_1\right);\ldots;v_{DEV\to IND}\left(C_N\right)\right\} \tag{198}$$

$$\tau_{IND}^{fnec}(\omega) = LOWA_{\omega}\{MIN(v_{DEV\to IND}(C_1), \tau_{DEV\to IND}(C_1));...;MIN(v_{DEV\to IND}(C_N), \tau_{DEV\to IND}(C_N))\} \quad (199)$$

Braumoeller's and Huth's results are reproduced in Table 3 for the sake of comparing with those obtained using LFLA. From Table 3, we see that LFLA for (ω = 0.7) agrees with Braumoeller's finding that the Balance of Military Force (MBF) is a strongly contributing factor in contradiction with Huth's original estimation. Note that the LFLA produces the same truth levels for all independent variables. This is to be expected as I have assumed that all input data had a PPP (perfect) degree of truth.

Table 3 Comparing results obtained by Huth, Braumoeller and LFLA (using Huth data for nuances and assuming PPP empirical confidence level for Huth's input variables).

Variable	Symbol	Huth Coefficient	p	Braumoeller Coefficient	p	LFLA (ω ~ 0.7) $v_{Dev\to Ind}$	$\tau_{Dev\to Ind}$
Strategic Location of Territory	SLT	2.637	.000	3.826	.000	NNN	LLL
Ties to Bordering Minority	TBM	– 0.073	.477	–0.920	.000	NNN	LLL
Political Unification	PUN	1.085	.000	–0.570	.007	NNN	LLL
Economic Value of Territory	EVT[5]	0.563	.000	0.702	.000	NNN	LLL
	EVT[6]			–0.022	.909		
Balance of Military Force	**MBF**	**0.083**	**.624**	**1.310**	**.007**	**VHH**	**LLL**
Prior Gain of Territory	PGT	–0.969	.000	–1.791	.000	NNN	LLL
Shared Alliance	SAL	–2.342	.000	–1.173	.000	NNN	LLL
Previous Settlement	PST	–3.339	.000	–18.87	.915	NNN	LLL
Prior Unresolved Dispute	PUD	3.761	.000	8.346	.000	NNN	LLL
Prior Loss of Territory	PLT	2.231	.000	5.039	.000	NNN	LLL
Decolonization Norm	DNO	0.740	.000	3.500	.000	NNN	LLL

Changing the degree of orness as shown in Table 4 does not change the conclusion that only MBF is a strongly contributing factor to causation. What indeed does change is the degree of truth for all independent variables, reaching the highest value of MMM (moderate) for ω ~ 0.5, that is, in the situation where we have an average – a midpoint – between full falsity and full truth with no means of opting either way. This position is a maximum since the truth level of all causal relations between each of the independent variables and the dependent variable diminish on either side of ω = 0.5, reaching a minimum at ω ~ 0.1 and ω ~ 0.9 of VLL (very low).

The results for LFLA multi-path causation (using equations[189]-[195]) are shown in Table 5 for various degrees of orness. In Table 5 I assume a perfect level of empirical confidence in Huth's input data for all variables and vary the level of orness (ω ~ 0.01, 0.1, 0.3, 0.5, 0.7, 0.9, 0.99). Table 5 shows the results for the degrees of nuance and truth at the three levels of variable combination that

[5] This is when EVT is considered as a contribution to systemic factors.
[6] This is when EVT is considered as a contribution to domestic factors.

Table 4 LFLA evaluation of multi-path causality for various levels of orness (ω ~ 0.1, 0.3, 0.5, 0.7, 0.9) using Huth data for nuances and assuming perfect empirical confidence

	ω ~ 0.1		ω ~ 0.3		ω ~ 0.5		ω ~ 0.7		ω ~ 0.9	
IND	ν	τ	ν	τ	ν	τ	ν	τ	ν	τ
SAL	NNN	VLL	NNN	LLL	NNN	MMM	NNN	LLL	NNN	VLL
SLT	NNN	VLL	NNN	LLL	NNN	MMM	NNN	LLL	NNN	VLL
PUN	NNN	VLL	NNN	LLL	NNN	MMM	NNN	LLL	NNN	VLL
TBM	NNN	VLL	NNN	LLL	NNN	MMM	NNN	LLL	NNN	VLL
DNO	NNN	VLL	NNN	LLL	NNN	MMM	NNN	LLL	NNN	VLL
PST	NNN	VLL	NNN	LLL	NNN	MMM	NNN	LLL	NNN	VLL
PUD	NNN	VLL	NNN	LLL	NNN	MMM	NNN	LLL	NNN	VLL
EVT	NNN	VLL	NNN	LLL	NNN	MMM	NNN	LLL	NNN	VLL
MBF	**VHH**	**VLL**	**VHH**	**LLL**	**VHH**	**MMM**	**VHH**	**LLL**	**VHH**	**VLL**
PGT	NNN	VLL	NNN	LLL	NNN	MMM	NNN	LLL	NNN	VLL
PLT	NNN	VLL	NNN	LLL	NNN	MMM	NNN	LLL	NNN	VLL

Table 5 LFLA evaluation of multi-path causation for various levels of orness (ω ~ 0.01, 0.1, 0.3, 0.5, 0.7, 0.9, 0.99) using Huth's input data for nuances and assuming perfect empirical confidence levels for Huth's input variables.

ω ~	ν_{SIC}	τ_{SIC}	ν_{DIC}	τ_{DIC}	ν_{MBF}	τ_{MBF}	ν_{MPC}	τ_{MPC}
0.01	NNN	NNN	NNN	NNN	VHH	NNN	VHH	PPP
0.10	NNN	NNN	NNN	NNN	VHH	VLL	VHH	VHH
0.30	VLL	NNN	VLL	NNN	VHH	LLL	VHH	HHH
0.50	MLL	VLL	MLL	VLL	VHH	MMM	HHH	MHH
0.70	MHH	LLL	MHH	LLL	VHH	LLL	MHH	HHH
0.80	VHH	VLL	VHH	VLL	VHH	LLL	HHH	VHH
0.90	PPP	VLL	PPP	VLL	VHH	VLL	HHH	VHH
0.99	PPP	NNN	PPP	NNN	VHH	NNN	VHH	PPP

Braumoeller considers – the systemic $(\nu_{SIC};\tau_{SIC})$, domestic $(\nu_{DIC};\tau_{DIC})$, and overall multi-path causation $(\nu_{MPC};\tau_{MPC})$. It also displays the values obtained for MBF (military balance of force).

Table 5 shows how the SIC and DIC combinations of independent variables contribute to multi-path causation. LFLA allows us to trace how causation progresses through the various paths. Several features are discernible from Table 5. First, at all levels of orness the systemic and domestic levels of causation have the same levels of nuance and truth. Second, the level of nuanced necessary causation reaches a perfect level (PPP) for both systemic and domestic combinations of independent variables for high values of orness (ω ≥0.90) and reaches a null level (NNN) for low (ω ≤0.10) or very high (ω ~1) values of orness. Generally, the level of nuanced necessary causation varies proportionally with the values of orness. High values of orness (as explicated in Chapter 2) correspond to an intersection of the fuzzy sets of all linguistic labels, and hence such combinations pick up the smaller (if not smallest) common set, which in this case will be the highest label. Very small values of orness correspond to a union of the fuzzy sets of all linguistic labels, and hence such combinations picks up the largest common set, which in this case will be the lowest label – the lowest label is necessarily implied in all subsequent labels. Hence, a causal combination will be an *orlike* operation with orness > 0.5, that is, behaving much more like crisp substitutability. On the other hand, a causal combination will be an *andlike* operation with orness < 0.5 – that is, behaving much more like crisp conjectural causation.

Third, the levels of truth (or confidence) also vary with the value of orness, if differently from the levels of nuanced causation. From Table 5, we see that the highest value of truth attained is LLL (low) and occurs for $\omega \sim 0.70$ which makes MHH (moderately high) causation the more likely (with the more confidence) to occur. Because $\omega \sim 0.70$ we can say that for both the systemic and domestic combinations (of independent variables) causation is behaving moderate-highly like substitution on these two paths. However, it is very unlikely to be either a crisp-like substitution (confidence level = NNN for $\omega \sim 0.99$) or a crisp-like conjectural causation (confidence level = NNN for $\omega \leq 0.30$). Fourth, multi-path causation occurs with a very high nuance level at very low ($\omega \sim 0.01$) or very high ($\omega \sim 0.99$) levels of orness – that is, full multi-path causation of either crisp-like conjectural type or crisp-like substitutability type is achievable and with a perfect level of confidence (confidence level = PPP for either $\omega \sim 0.01$ or $\omega \sim 0.99$). The nuance level of multi-path causation decreases for ω converging toward the mid-value 0.5 from either side. In other words, a fuzzy combination of conjectural-like and substitutability-like multi-path causation is still possible with HHH (high) to MHH (moderately high) degree of fuzziness with a confidence level of MHH to HHH for values of orness near the mid-point. It is to be noted that these results depend

Table 6 LFLA evaluation of multi-path causality for orness level $\omega \sim 0.7$ using Huth data for nuances and assuming uniformly random truth values for all variables

$\omega \sim 0.7$ with Uniformly Randomized Input Truth Values for Independent and Dependent Variables					
ν_{SIC}	τ_{SIC}	ν_{DIC}	τ_{DIC}	ν_{MPC}	τ_{MPC}
PPP	PPP	PPP	PPP	PPP	PPP

Ind. Var.	$\nu_{Dep \to Ind}$	$\tau_{Dep \to Ind}$
SAL	VHH	PPP
SLT	VHH	PPP
PUN	VHH	PPP
TBM	VHH	PPP
DNO	VHH	PPP
PST	VHH	PPP
PUD	VHH	PPP
EVT	VHH	PPP
MBF	VHH	PPP
PGT	VHH	PPP
PLT	VHH	PPP

on the assumption that the input data was assumed to have a perfect confidence level (PPP). For the sake of comparison I present in Table 6 the results for $\omega \sim 0.7$ with (uniform) *randomly* distributed truth values for all input (dependent and independent) variables, but keeping the same empirical values for the nuance values of the variables.

It turns out that all the nuances values of the fuzzy necessity of all independent variables become VHH with perfect truth values. It also leads to perfect multi-path causation with perfect truth values for SIC (systemic), DIC (domestic), and MPC (multi-path causation). In this case, we have a perfect fuzzy combination of conjectural-like and substitutability-like causation (since $\omega \sim 0.7$) at a perfect level of confidence. This suggests that the missing data on truth values (or, equivalently, confidence levels) of the empirical data is an extremely important factor in the analysis of causality. Determining the truth values of the independent and dependent variables is as important as measuring the values (nuances) of these variables.

Appendix A: Fuzzification Scheme

The trapezoidal membership function used to fuzzify Huth's data is defined in equation [200] where the constants a, b, c, and d are listed in Table A.1. Figure A.1 shows the shape of the nine linguistic terms {NNN, VLL, LLL, MLL, MMM, MHH, HHH, VHH, PPP} in the trapezoidal representation.

$$\mu(x,a,b,c,d) = \begin{cases} 0 & x < a, x > d \\ \dfrac{x-a}{b-a} & a \le x \le b \\ 1 & b < x < c \\ \dfrac{d-x}{d-c} & c \le x \le d \end{cases} \tag{200}$$

Table A.1 Scheme used to convert Huth's (1996) numerical data to linguistic data

a	b	c	d	Linguistic Value	Meaning
0.0000	0.0000	0. 0125	0.1125	NNN	Null
0.0125	0.1125	0.1375	0.2375	VLL	Very Low
0.1375	0.2375	0.2625	0.3625	LLL	Low
0.2625	0.3625	0.3875	0.4875	MLL	Moderately Low
0.3875	0.4875	0.5125	0.6125	MMM	Moderate
0.5125	0.6125	0.6375	0.7375	MHH	Moderately High
0.6375	0.7375	0.7625	0.8625	HHH	High
0.7625	0.8625	0.8875	0.9875	VHH	Very High
0.8875	0.9875	1.0000	1.0000	PPP	Perfect

Not surprisingly, when applied to Huth's original data a number of features clearly stand out. First, all of his variables except for MBF (military balance of force) assume dichotomous values (either 0 or 1) and, hence, as expected, correspondingly convert either to NNN or PPP. For the variable MBF, those of its values that are either 0 or 1 are also converted into NNN and PPP, respectively. For all other values, in most cases I find that a specific crisp value MBF, with 0 < value(MBF) < 1, there are two fuzzy sets to which Huth's crisp MBF value belongs, but with different membership degrees μ.

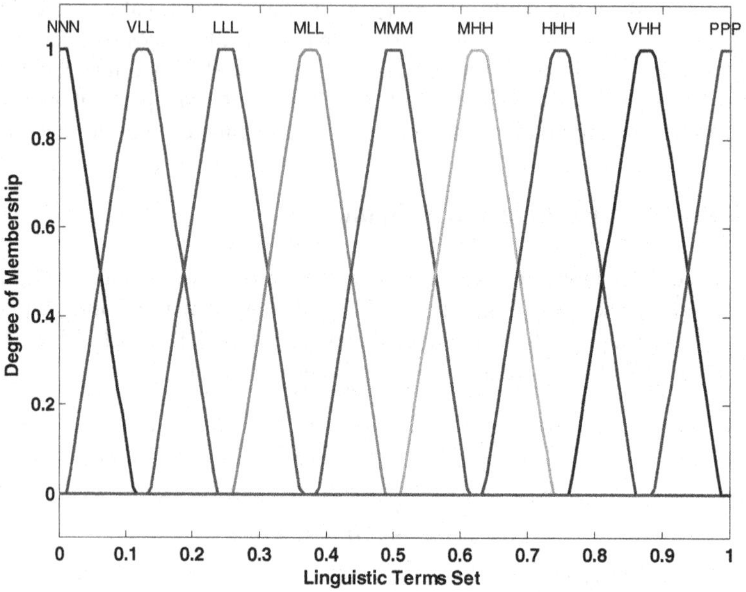

Fig. A.1 Trapezoidal Representation of 9 Linguistic Terms Set

As an illustration let us consider Huth's original crisp value MBF = 0.6399999. I find that this corresponds to two contiguous linguistic categories MHH and HHH with corresponding degrees of membership μ_1=0.975 and μ_2=0.025 – that is, 0.6399999 belongs to MHH fuzzy set to a .975 degree and to HHH with a .025 degree. I follow the criterion of choosing the category corresponding to the higher degree of membership as the most representative fuzzy value of the specific crisp value measured by Huth. This way of fuzzifying Huth's original crisp data hence depends on a somewhat arbitrary choice but this is not a problem for three reasons. First, I am put into the difficult task of testing an LFLA approach with a crisp data which was not meant to be used with fuzzy logic. As already pointed out in the main text, Huth's data lacks the truth values (confidence levels) of the measured data, thereby making its use in testing LFLA quite limited. The proposed scheme for fuzzifying Huth's data partially resolves this difficulty. Second, this fuzzification scheme is really meant as an illustration on how one might go by using the same crisp data to compare LFLA to methods that are designed to work with crisp data in the first place. I am sure that the reverse story will face the crisp methodologist with the same dilemma – that is, imagine that someone using a crisp method decides to test it using fuzzy data. This person would be obliged to de-fuzzify the fuzzy data to use it in his/her methodology, thereby being forced to make an arbitrary choice during the de-fuzzication procedure.[7] There are many ways to carry out such a task (artificial intelligence literature is extremely rich on this respect). Third, in LFLA membership functions do not intervene at all in LFLA – hence, the arbitrariness ensuing from their determination and manipulation is altogether absent. Having said this, the scheme that I use, despite all the problems, allows me systematically to do the fuzzification of an originally crisp data. Table A.2 displays a sample (only a sample since Huth's data has 8328 points for each variable) of the original and converted data.

Table A.2 Fuzzification of a Sample of Huth's Data

DEV		SAL		SLT		PUN		TBM		DNO		PST		PUD		EVT		MBF		PGT		PLT	
1	PPP	1	PPP	1	PPP	1	PPP	0	NNN	0	NNN	0	NNN	1	PPP	1	PPP	0.629999995	MHH	0	NNN	1	PPP
1	PPP	1	PPP	1	PPP	1	PPP	0	NNN	0	NNN	0	NNN	1	PPP	1	PPP	0.699999988	HHH	0	NNN	1	PPP
1	PPP	1	PPP	1	PPP	1	PPP	0	NNN	0	NNN	0	NNN	1	PPP	1	PPP	0.639999986	MHH	0	NNN	1	PPP
1	PPP	1	PPP	1	PPP	1	PPP	0	NNN	0	NNN	0	NNN	1	PPP	1	PPP	0.550000012	MMM	0	NNN	1	PPP
1	PPP	1	PPP	1	PPP	1	PPP	0	NNN	0	NNN	0	NNN	1	PPP	1	PPP	0.699999988	HHH	0	NNN	1	PPP
1	PPP	1	PPP	1	PPP	1	PPP	0	NNN	0	NNN	0	NNN	1	PPP	1	PPP	0.620000005	MHH	0	NNN	1	PPP
1	PPP	1	PPP	1	PPP	1	PPP	0	NNN	0	NNN	0	NNN	1	PPP	1	PPP	0.579999983	MHH	0	NNN	1	PPP
1	PPP	1	PPP	1	PPP	1	PPP	0	NNN	0	NNN	0	NNN	1	PPP	1	PPP	0.649999976	MHH	0	NNN	1	PPP
1	PPP	1	PPP	1	PPP	1	PPP	0	NNN	0	NNN	0	NNN	1	PPP	1	PPP	0.620000005	MHH	0	NNN	1	PPP
1	PPP	1	PPP	1	PPP	1	PPP	0	NNN	0	NNN	0	NNN	1	PPP	1	PPP	0.660000026	MHH	0	NNN	1	PPP

[7] I would dare to say that most of us do this anyway when using numbers to measure "real-life" data but do no account for it – hence, one of the main reasons for advocating methods based on linguistic fuzzy logic.

Chapter 8
Conclusion

This book proposes a linguistic fuzzy-logic approach – LFLA – to analyze the process of fuzzy decision making, fuzzy strategic interaction and fuzzy game theory, causality using fuzzy propositional logic, and fuzzy data analysis. Much scholarship has been produced on these issues area of decision making in political science and social sciences in general. Much has been achieved in this regard. This book contributes to this body of literature with a new way of looking at these problems.

Much of what counts as social science today is based on a Boolean logic with two truth values, 0 and 1. While this assumption – because it is really an assumption, if in most situations not even acknowledged as such – has proved to be very useful in the production of systematic knowledge, it is neither theoretically nor empirically the only possible logic. The study of logic is a vast discipline of knowledge in itself and the diversity of possible logics that scholars study is very rich. However, except for a very limited number of studies, by and large social science practitioners take it for granted that Boolean logic is THE logic that should underlie social science inquiry. The task set out for this book was how to go beyond this assumption by showing that linguistic fuzzy logic can be very useful in analyzing empirical data and studying causation in social science theories. Linguistic fuzzy logic has a greater expressive power which allows us to effectively deal with meanings of words and phrases expressed in natural language. This thus enables us to express and analyze the role of irreducible uncertainties and vagueness assuming various manifestations in validating our theories about the human world, thereby allowing us to capture more accurately human reasoning, cognition, and communication.

A key contribution of linguistic fuzzy logic approach to the study of causation and other processes in social sciences is to emphasize the need not only of measuring the linguistic values of both independent and dependent variables (termed in this book as nuances), a call that regular fuzzy-set theoretic approach already make. LFLA also shows the crucial role of a dimension of both independent and dependent variables which by and large has been completely ignored in measuring and analyzing the empirical data – the truth value of the nuance values (with the latter commonly termed simply as 'the values") of the variables and combinations thereof. Hence, LFLA analysis is more demanding in terms of empirical data – it requires not only a fuzzification of the values of categorical variables but also an estimation of our confidence in the evaluation of these fuzzified values (termed equivalently truth or confidence levels). As shown in the various illustrations

B. Arfi: Linguistic Fuzzy Logic Methods in Social Sciences, STUDFUZZ 253, pp. 175–179.
springerlink.com

considered in this book, different values of the truth levels shape the final conclusions of the analysis in important ways. A comprehensive and systematic analysis of data and causation (and other processes) cannot ignore this crucial aspect of the analysis.

This book is, to the best of my knowledge, a first in its kind in suggesting a linguistic fuzzy-logic methodology in social sciences. There are no other books which might be considered as direct competitors. There are however two books (published by social scientists) of a similar nature with which this book can be more or less compared. This book is a natural continuation of the "logic" and intent of these two books in the sense that it takes off where they stop in introducing fuzzy logic methods to social sciences. The "edge" of this book is that it advocates fuzziness to the "full extent," by "computing with words," not numbers.

The first book is by Charles Ragin, *Fuzzy-Set Social Science* (University of Chicago Press, 2000). This book was one of the best attempts to introduce fuzzy-set methods to social sciences in way that addresses two related problems in then existing methodologies: first, the divide between quantitative and qualitative methods and, second, the problem of the "homogenizing assumptions" about cases and causes. Ragin's innovative contribution consists in extending diversity-oriented research strategies which provide a powerful connection between theory and data analysis. This book does however suffer from two main weaknesses due to its reliance on "numbers." First is the problem of evaluating the membership functions of elements of the fuzzy sets. There are no first-principled ways to do that and hence this brings in an ineradicable element of arbitrariness that cannot be systematically accounted for or measured. Second is the reliance on statistical methods such as probability benchmarking to complement the fuzzy-sets method, thereby making the approach still reliant on conventional statistical analyses which are developed for crisp-sets logic, not fuzzy-sets logic.

The second book is by Michael J. Smithson and Jay Verkuilen, *Fuzzy Set Theory: Applications in the Social Sciences* (Sage, 2006). First, this book provides an accessible introduction to fuzzy set theory which specifically targets its applicability to the social sciences. There are many introductory books on fuzzy-set theory on the market. Yet all of them have been written for engineering audiences, thereby lacking the illustrations which are necessary for the non-initiated social scientists to grasp these methods. Second, this book improves Ragin's book approach to data analysis by presenting a systematic and practical guide for social scientists to combine fuzzy set theory with standard statistical techniques and model-testing. Although this book is a good and systematic improvement on Ragin's book, it is still plagued with the same problems related with the numerical nature of the membership functions, even if it is much more explicit about these issues.

Three fundamental features make the present book unique compared to these two books and similar works published in social sciences methodology. First, all published works are interested in using numerical fuzzy-sets theoretic tools as the basis for new methodologies for empirical analysis. My approach seeks to analyze social science phenomena using tools developed based on linguistic fuzzy logic. Second, all published works are based on the use of what is known as membership function or degree of belongingness to sets, a number between zero and one. My approach computes with words all the way down and does not use membership functions. I instead consider the degrees of belongingness to a category as a linguistic variable termed as a nuance value. I also introduce a second type of fuzziness at a deeper level than what extant works consider. The membership degrees are also fuzzy sets, or more accurately, linguistically expressed truth values of the nuance values.

Third, the book contributes to shed light on lingering disputes on tradeoffs between rigor in reasoning, precision in conceptualization, correspondence with empirical reality, and logical consistency. Most of what counts as mainstream social science today assumes that vagueness is an epistemic problem that can be reduced using increasingly more sophisticated statistical, formal and other tools in analyzing the empirical world. The linguistic fuzzy logic approach is built on the premise that vagueness is essentially not epistemic. Vagueness is inherent to the constitutive and causal logic underlying the phenomena themselves. From this perspective, using Boolean two-valued logic as a means of, for example, systematically studying the consistency and coherence of theories might very well distort the very logic of the phenomenon under study. Our minds are language-shaped and as such the logic of the workings of our minds is/should be language shaped. We do not live in a binary world of 0s and 1s. We live in a world the logic of which is best understood using vague linguistic expressions, not crisp numbers. Our tools of analysis should hence reflect this aspect of the human condition. And this is what the book seeks to do.

Fourth, this book develops a new kind of game theory – termed as LFL game theory – which departs from conventional (crisp) game theory by being based on linguistic fuzzy logic rather than Boolean two-valued logic. Going from a Boolean logic to LFL lifts the PD dilemma and allows the emergence of a strong Nash equilibrium. Moreover, there always exists at least one Pareto-optimum Nash equilibrium. A second major departure of this work is the use of "words to compute" the solutions of the game. The vagueness that inherently exists in real strategic situations is preserved in LFL games. It is not considered an epistemic issue but rather as an ontological issue and is hence incorporated in the very logic underpinning the reasoning process. A third departure has to do with the extent of stylization in analyzing real situations. Although one cannot avoid a certain level of stylization in formalizing strategic interactions between actors, the level of stylization in LFL games is much less than in conventional game theory. This makes LFL game theory much more straightforwardly relevant to analyzing real-world situations of strategic interaction.

Assuming that the logic underlying social sciences is of a linguistic fuzzy type has important implications for research in social sciences. First, much ink and talk have been spent on the qualitative-quantitative divide in social sciences, thereby contributing to enrich our approaches and knowledge about the world. However, much of this debate does not go far enough from the perspective of this book. Although logic informs all reasoning, whether it is based on quantitative inquiry or qualitative work, many researchers are not fully aware of the implications and limitations of the logic underlying their inquiry and how this logic informs the epistemology which supports their empirical work. Linguistic fuzzy logic approach is one important step toward bridging the gap.

Second, both qualitative and quantitative research engage in data reduction, with the former doing the reduction through using words, categories, and themes and the latter doing the reduction through numerical and often statistical tools. The analysis of data (after reduction) is done in qualitative research through categorizing and comparing and in quantitative methods through statistical inference and estimation of likelihood and the like. Of course, these differences are matters of degree. However, generally speaking, quantitative research is believed by its practitioners to generate reliable population-based and generalizable data and is hence well suited for studying cause-and-effect relationships. The fact that qualitative data typically involves words and quantitative data involves numbers prompts many researchers to believe that the latter is better and more scientific than the former. The linguistic fuzzy logic approach proposed in this book is one way to create a bridge between the qualitative and quantitative sides of this debate. The methodology of this book is both rigorous and incorporates vagueness at its very heart. It proposes mathematically rigorous algorithms to reduce and analyze data which is expressed in a natural language, hence preserving the vague meanings of linguistically defined information.

Third, a key contribution of linguistic fuzzy logic approach to the study of causation and other processes in social sciences is to emphasize the need not only of measuring the linguistic values of both independent and dependent variables, a call that numerical fuzzy-set theoretic approach already makes (e.g., Ragin's book). The approach of the book also shows the crucial role of a dimension of both independent and dependent variables, which by and large has been completely ignored in measuring and analyzing the empirical data, namely, the truth value of the linguistic values of the variables. Hence, linguistic fuzzy-logic analysis is more demanding in terms of empirical data because it requires not only a fuzzification of the values of categorical variables but also an estimation of our confidence in the evaluation of these fuzzified values. As shown in the various illustrations considered in this book, different values of the truth or confidence levels do shape the final conclusions of the analysis. A comprehensive and systematic analysis of data and causation and other processes cannot ignore this crucial aspect of the analysis. The concept of linguistic truth level, which naturally emerges in linguistic fuzzy logic, makes this methodology more or less self-contained.

Fourth, linguistic fuzzy-logic methodology enriches our very notion of causation itself. Conventional practice and conceptualization discuss causation in terms of necessity, sufficiency, and combinations thereof. This book shows that such notions are too restrictive given the richness of natural language which is constitutive of social reality as we construct it, experience it, and analyze it. Instead of talking about necessity and sufficiency, as most of what counts as social science practice does, linguistic fuzzy-logic methodology offers to us novel notions of fuzzy causation. Hence, for example, instead of speaking of crisp necessity we should speak of fuzzy necessity, which is a concatenation of necessity and contingency, a notion quite different from probabilistic necessity which is essentially still based on crisp notion of necessity. In linguistic fuzzy-logic, crisp necessity and crisp contingency become limiting cases of fuzzy necessity.

Fifth, the approach of the book helps improve the connection between the scholarly and policy worlds. Although there are many reasons why such a linkage does not always evolve in mutually beneficial ways, one impediment that seems to reinforce the lack of effective communication is that the media used by scholars in their research are often times very opaque to the uninitiated practitioners. Conversely, scholars must often play the role of translators between the policy world and their theories about social and political reality. The methodology proposed in this book lends itself quite naturally to provide an effective means of communication between scholars and practitioners since the approach greatly facilitates the exchange and analysis of "raw" linguistically expressed information. Scholars and their theories and hypotheses can directly "speak" to the policy world and incorporate "raw elements" of the policy world in their theories and hypotheses.

References

Alexander, P.J.: Entropy and Popular Culture. American Sociological Review 61(1), 171–174 (1996)

Arfi, B.: "Spontaneous" Interethnic Order: The Emergence of Collective, Path Dependent Cooperation. International Studies Quarterly 44(4), 563–590 (2000)

Arfi, B.: Fuzzy Decision Making in Politics: A Linguistic Fuzzy-Set Approach (LFSA). Political Analysis 13, 23–56 (2005)

Billot, A.: Economic Theory of Fuzzy Equilibria: An Axiomatic Analysis. Springer, New York (1992)

Bolton, J.R.: The Risks and Weaknesses of the International Criminal Court from America's Perspective. Law and Contemporary Problems 64(1), 167–180 (2001)

Bordogna, G., Passi, G.: A Fuzzy Linguistic Approach Generalizing Boolean Information Retrieval: A Model and its Evaluation. Journal of American Society of Information Sciences 44, 70–82 (1993)

Borges, P.S.S., Pacheco, R.C.S., Barcia, R.M., Khator, S.K.: A Fuzzy Approach to the Prisoner's Dilemma. Biosystems 41, 127–137 (1997)

Braumoeller, B.F.: Causal Complexity and the Study of Politics. Political Analysis 11, 209–233 (2003)

Brown, M.E., Cote Jr., O.R., Lynn-Jones, S.M., Miller, S.E. (eds.): Rational Choice and Security Studies: Stephen Walt and His Critics. The MIT Press, Cambridge (2000)

Butnariu, D.: Fuzzy Games: A Description of the Concept. Fuzzy Sets and Systems 1, 181–192 (1978)

Chan, S.: Explaining War Termination: a Boolean Analysis of Causes. Journal of Peace Research 40(1), 49–66 (2003)

Cioffi-Revilla, C.A.: Fuzzy Sets and Models of International Relations. American Journal of Political Science 25(1), 129–159 (1981)

Cover, T.M., Thomas, J.A.: Elements of Information Theory. John Wiley and Sons, New York (1991)

De Wilde, P.: Fuzzy Utility and Equilibria. IEEE Transactions on Systems, Man, and Cybernetics—Part B: Cybernetics 34(4), 1774–1785 (2004)

Delgado, M., Herrera, F., Herrera-Viedman, E., Martinez, L.: Combining Numerical and Linguistic Information in Group Decision Making. Journal of Information Sciences 107, 177–194 (1998)

Delgado, M., Verdegay, J.L., Vila, M.A.: On Aggregation Operations of Linguistic Labels. International Journal of Intelligent Systems 8, 351–370 (1993)

Dutta, B.: Fuzzy Preferences and Social Choice. Mathematical Social Sciences 13, 215–229 (1987)

Fearon, J.D., Laitin, D.: Explaining Interethnic Cooperation. American Political Science Review 90(4), 715–735 (1996)

Filev, D., Yager, R.J.: Analytic Properties of Maximum Entropy OWA Operators. Information Sciences 85, 11–25 (1995)

Fuller, R., Majlender, P.: An Analytic Approach for Obtaining Maximal Entropy OWA Operator Weights. Fuzzy Sets and Systems 124, 53–57 (2001)

Geslin, S., Salles, M., Ziad, A.: Fuzzy Aggregation in Economic Environments: I. Quantitative Fuzziness, Public Goods and Monotonicity Assumptions. Mathematical Social Sciences 45, 155–166 (2003)

Gill, J.: The Political Entropy of Vote Choice: An Empirical Test of Uncertainty Reduction. In: Prepared for the 1997 Annual Meeting of the American Political Science Association, Washington, DC, August 27-31 (1997)

Goertz, G., Starr, H. (eds.): Necessary Conditions: Theory, Methodology, and Applications. Rowman & Littlefield Publishers, Inc., New York (2003)

Goertz, G., Mahoney, J.: Two-Level Theories and Fuzzy-Set Analysis. Sociological Methods & Research 33(4), 497–538 (2005)

Goertz, G.: Cause, Correlation, and Necessary Conditions. In: Goertz, G., Starr, H. (eds.) Necessary Conditions: Theory, Methodology, and Applications, pp. 47–64. Rowman & Littlefield Publishers Inc., New York (2003a)

Goertz, G.: The Substantive Importance of Necessary Condition Hypotheses. In: Goertz, G., Starr, H. (eds.) Necessary Conditions: Theory, Methodology, and Applications, pp. 65–04. Rowman & Littlefield Publishers Inc., New York (2003b)

Goertz, G.: Constraints, Compromises, and Decision Making. The Journal of Conflict Resolution 48(1), 14–37 (2004)

Golan, A., Judge, G., Karp, L.: A Maximum Entropy Approach to Estimation and Inference in Dynamic Models: Counting Fish in the Sea Using Maximum Entropy. Journal of Economic Dynamics and Control 20, 559–582 (1996)

Gurocak, E.R., Whittlesey, N.K.: Multiple Criteria Decision Making: A Case Study of the Columbia River Salmon Recovery Plan. Environmental and Resource Economics 12, 479–495 (1998)

Hajek, P.: Mathematics of Fuzzy Logics. Kluwer Academic Publishers, Boston (1998)

Herrera, F., Herrera-Viedman, E.: Aggregation Operations for Linguistic Weighted Information. IEEE Transactions on Systems, Man, and Cybernetics – Part A: Systems and Humans 27(5), 646–656 (1997)

Herrera, F., Herrera-Viedma, E.: Linguistic Decision Analysis: Steps for Solving Decision Problems under Linguistic Information. Fuzzy Sets and Systems 115, 67–82 (1998)

Herrera, F., Herrera-Viedman, E.: Linguistic Decision Analysis: Steps for Solving Decision Problems under Linguistic Information. Fuzzy Sets and Systems 115, 67–82 (2000)

Herrera, F., Herrera-Viedman, E.: Linguistic Decision Analysis: Steps for Solving Decision Problems under Linguistic Information. Fuzzy Sets and Systems 115, 67–82 (2000)

Herrera, F., Martinez, L.: A 2-Tuple Fuzzy Linguistic Representation Model for Computing with Words. IEEE Transactions on Fuzzy Systems 8(6), 746–752 (2000)

Herrera, F., Herrera-Viedma, E., Verdegay, J.L.: Direct Approach Processes in Group Decision Making using Linguistic OWA Operators. Fuzzy Sets and Systems 79, 175–190 (1996)

Herrera, F., Herrera-Viedma, E., Verdegay, J.L.: Direct Approach Processes in Group Decision Making using Linguistic OWA Operators. Fuzzy Sets and Systems 79, 175–190 (1996)

Herrera, F., Herrera-Viedma, E., Martinez, L.: A Fusion Approach for Managing Multi-granularity Linguistic Term Sets in Decision Making. Fuzzy Sets and Systems 114, 43–58 (2000)

Herrera, F., Lopez, E., Rodriguez, M.A.: A Linguistic Decision Model for Promotion Mix Management Solved with Genetic Algorithms. Fuzzy Sets and Systems (2002) (in press)

Herrera, F., Lopez, E., Mendena, C., Rodriguez, M.A.: A Linguistic Decision Model for Personnel Management Solved with a Linguistic Biobjective Genetic Algorithm. Fuzzy Sets and Systems 118, 47–64 (2001)

Herrera-Viedma, E.: Modeling the Retrieval Process for an Information Retrieval System Using an Ordinal Fuzzy Linguistic Approach. Journal of American Society of Information Sciences 52(6), 460–475 (2001)

Heyting, A.: Intuitionism: An introduction. North-Holland, Amsterdam (1971)

Ho, N.C., Khang, T.D.: Hedge Algebras, Linguistic-Valued Logic and their Application to Fuzzy reasoning. International Journal of Uncertainty, Fuzziness and Knowledge-Based Systems 7(4), 347–361 (1999)

Ho, N.C., Nam, H.V.: An Algebraic Approach to Linguistic Hedges in Zadeh's Fuzzy Logic. Fuzzy Sets and Systems 129, 229–254 (2002)

Huth, P.K.: Standing Your Ground: Territorial Disputes and International Conflict. The University of Michigan Press, Ann Arbor (1996)

Jaynes, E.T.: Information Theory and Statistical Mechanics. Physical Reviews 106, 620–630; 108, 171–190 (1957)

Jaynes, E.T.: On the Rationale of Maximum-Entropy Methods. Proceedings of the IEEE 70(9), 939–952 (1982)

Jaynes, E.T.: Clearing up Mysteries. In: Skilling, J. (ed.) 'The original Goal' in Maximum Entropy and Bayesian Methods. Kluwer, Dordrecht (1989)

Kim, W.K., Lee, K.H.: Generalized Fuzzy Games and Fuzzy Equilibria. Fuzzy Sets and Systems 122, 293–301 (2001)

Klir, G.J., Yuan, B.: Fuzzy Sets and Fuzzy Logic: Theory and Applications. PrenticeHall, NJ (1995)

Kydd, A.: Trust, Reassurance, and Cooperation. International Organization 54(2), 325–357 (2000)

Lake, D.: Entangling Relations: American Foreign Policy in Its Century. Princeton University Press, Princeton (1999)

Layder, D.: Beyond Empiricism? The Promise of Realism. Philosophy of the Social Sciences 15(3), 255–274 (1985)

Li, K.W., Karray, F., Hipel, K.W., Marc Kilgour, D.: Fuzzy Approaches to the Game of Chicken. IEEE Transactions on Fuzzy Systems 9(4), 608–623 (2001)

Maeda, T.: Characterization of the Equilibrium Strategy of the Bimatrix Game with Fuzzy Payoff. Journal of Mathematical Analysis and Applications 251, 885–896 (2000)

Maeda, T.: On Characterization of Equilibrium Strategy of Two-Person Zero-Sum Games with Fuzzy Payoffs. Fuzzy Sets and Systems 139, 283–296 (2003)

Mendel, J.M.: Uncertain Rule-Based Fuzzy Logic Systems. Prentice Hall, Upper Saddle River (2001)

Mintz, A., Geva, N.: The Poliheuristic Theory of Foreign Policy Decision-making. In: Geva, N., Mintz, A. (eds.) Decision Making on War and Peace: The Cognitive-Rational Debate. Lynne Rienner, Boulder (1997)

Mintz, A., Geva, N., DeRouen, K.: Mathematical Models of Foreign Policy Decision-making: Compensatory vs. Noncompensatory. Synthese 100, 441–460 (1994)

Mintz, A.: The Decision to Attack Iraq: A Noncompensatory Theory of Decision-making. Journal of Conflict Resolution 3, 595–618 (1993)

Mintz, A.: The Noncompensatory Principle of Coalition Formation. Journal of Theoretical Politics 7, 335–349 (1995)

Neyman, A.: Strategic Entropy and Complexity in Repeated Games. Games and Economic Behavior 29, 191–223 (1999)

Novak, V., Perfilieva, I., Mockor, J.: Mathematical Principles of Fuzzy Logic. Kluwer Academic Publishers, Boston (1999)

Nurmi, H.: Voting Paradoxes and Referenda. Social Choice and Welfare 15, 333–350 (1998)

Nurmi, H., Meskanen, T.: Voting Paradoxes and MCDM. Group Decision and Negotiation 9, 297–313 (2000)

O'Connell, T.C., Stearns, R.E.: On Finite Strategy Sets for Finitely Repeated Zero-sum Games. Games and Economic Behavior 43, 107–136 (2003)

Paris, J., Vencovskfi, A.: In Defense of the Maximum Entropy Inference Process. International Journal of Approximate Reasoning 17, 77–103 (1996)

Priest, G., Routley, R., Norman, J.: Paraconsistent Logic: Essays on the Inconsistent. Philosophia Verlag, Munich (1989)

Ragin, C.C.: The Comparative Method. University of California Press, Berkeley (1987)

Ragin, C.C.: Fuzzy-Set Social Science. University of Chicago Press, Chicago (2000)

Redd, S.B.: The Influence of Advisers on Foreign Policy Decision Making: An Experimental Study. The Journal of Conflict Resolution 48(1), 335–364 (2004)

Richardson, G.: The Structure of Fuzzy Preferences: Social Choice Implications. Social Choice Welfare 15, 359–369 (1998)

Sanjian, G.S.: Fuzzy Set Theory and US Arms Transfers: Modeling the Decision-Making Process. American Journal of Political Science 32(4), 1018–1046 (1988)

Sanjian, G.S.: Great Power Arms Transfers: Modeling the Decision-Making Processes of Hegemonic, Industrial, and Restrictive Exporters. International Studies Quarterly 35, 173–03 (1991)

Sanjian, G.S.: A Fuzzy Set Model of NATO Decision-Making: The Case of Short-Range Nuclear Forces in Europe. Journal of Peace Research 29(3), 271–286 (1992)

Shannon, C.E.: A Mathematical Theory of Communication. Bell System Technology Journal 27, 379-423, 623-56 (1948)

Shannon, C.E.: The Mathematical Theory of Communication. University of Illinois Press, Urbana (1949)

Shen, E.Z., Perloff, J.M.: Maximum Entropy and Bayesian Approaches to the Ratio Problem. Journal of Econometrics 104, 289–313 (2001)

Simon, H.A.: The Sciences of the Artificial, 3rd edn. MIT Press, Cambridge (1996)

Skocpol, T.: States and Social Revolutions: A Comparative Analysis of France, Russia, and. Cambridge University Press, Cambridge (1979)

Smithson, M.J.: Fuzzy Set Analysis for the Behavioral and Social Sciences. Springer, New York (1987)

Smithson, M.J., Verkuilen, J.: Fuzzy Set Theory: Applications in the Social Sciences. Sage, Thousand Oaks (2006)

Theil, H.: The Desired Political Entropy. American Political Science Review 63(2), 521–525 (1969)

Ullah, A.: Uses of Entropy and Divergence Measures for Evaluating Econometric Approximations and Inference. Journal of Econometrics 107, 313–326 (2002)

Yager, R.R.: On Ordered Weighted Averaging Aggregation Operators in Multicriteria Decision Making. IEEE Transactions on Systems, Man, and Cybernetics 18(1), 183–190 (1988)

Zadeh, L.A.: Decision-Making in a Fuzzy Environment. Management Science 17(4), B141–B164 (1970)

Zadeh, L.A.: From Computing with Numbers to Computing with Words – From Manipulation of Measurements to Manipulation of Perceptions. IEEE Transactions on Circuits and Systems – I: Fundamental Theory and Applications 45(1), 105–119 (1999)

Zinnes, D.A.: Constructing Political Logic: The Democratic Peace Puzzle. Journal of Conflict Resolution 48(3), 430–454 (2004)

Zinnes, D.A.: The democratic Peace Puzzle. Journal of Conflict Resolution 38(3), 430–454 (2004)